Radioactivity and Radiation

"Radiation Protection"

© by Claus Grupen

Claus Grupen · Mark Rodgers

Radioactivity and Radiation

What They Are, What They Do,
and How to Harness Them

 Springer

Claus Grupen
Department of Physics
University of Siegen
Siegen
Germany

Mark Rodgers
Bluesmith Information Systems
Leeds
UK

ISBN 978-3-319-82554-0   ISBN 978-3-319-42330-2   (eBook)
DOI 10.1007/978-3-319-42330-2

With cartoons by N. Downes, C. Grupen (partly inspired by his son Cornelius), L. Murchetz, V. Renčín, J. Wolter; copyright by the cartoonists. All line drawings, unless otherwise noted, are by C. Grupen, who retains the copyright for them.
Certain diagrams were produced by M. Rodgers, who retains the copyright for them, using Jaxodraw by D. Binosi and L. Theußl, http://dx.doi.org/10.1016/j.cpc.2004.05.001, and the GNU Image Manipulation Program, www.gimp.org.
Graphic design is by Armbrust Design, Stefan H. Armbrust, Siegen, Germany.

Printed on acid-free paper

This Springer imprint is published by Springer Nature
The registered company is Springer International Publishing AG Switzerland

Disclaimer

This book provides an overview of radiation physics. It is intended for members of the general public who are interested in radioactivity and radiation, and also science enthusiasts more generally. We have tried to keep the information correct and up-to-date. However, any reliance on such information is strictly at your own risk. This refers to the availability of the commercial products shown and also to those places where web sites are given. Web addresses can change quite rapidly, but we have made sure that the sites given in the text are available at the time of printing of this book. These web sites are not under our control. We have also tried to get written permission for the use of all pictures (relating to radiation detection, radiation instruments and other topics) which are shown in this book. Not in all cases did the companies providing such instruments or information respond to our request for permission to show their products. If any material in this book is being described or referred to improperly or incompletely, we would be grateful to be informed so that we can consider this for future editions of the book.

Claus Grupen
Mark Rodgers

Preface

Like taxes, radioactivity has long been with us and in increasing amounts; it is not to be hated and feared, but accepted and controlled. Radiation is dangerous, let there be no mistake about that – but the modern world abounds in dangerous substances and situations too numerous to mention, ... Consider radiation as something to be treated with respect, avoided when practicable, and accepted when inevitable.

Ralph Lapp 1917–2004

Radiation is everywhere. In this book, we are concerned with ionising radiation, i.e. radiation that can ionise ordinary atoms, such as the rays which come from radioactive decay. Almost everything is radioactive. Radiation emerges from the soil, it is in the air, and our planet is continuously being bombarded with energetic cosmic radiation. Even the human body is radioactive: about 9000 decays of unstable nuclei occur per second in a normal human body. In the early days of the Earth, when our planet formed from the debris of the early solar system, the radiation level was much higher. It is possible that the origin, development and biodiversity of life have been positively influenced by ionising radiation.

Since the early twentieth century, mankind has been able to artificially create radioactive nuclei: particularly so since the discovery of nuclear fission in the late 1930s. As early as 1905, Pierre Curie remarked that radium in the hands of criminals could be a disaster. Also, Louis de Broglie noted in his Nobel lecture in 1927 that he did not know whether science in the hands of humans is a good or a bad thing. The bombing of Hiroshima and Nagasaki in 1945 with nuclear weapons clearly demonstrated the disastrous effect of ionising radiation. The Nobel laureate for medicine, Maurice H.F. Wilkins, said contemplatively: "We have now reached the point where it is an open question as to whether doing more science is a good thing". The nuclear accidents near Harrisburg at the Three Mile Island reactor (1979), in Chernobyl (1986), Tokaimura (1999) and Fukushima (2011) clearly demonstrated that nuclear fission requires high-quality safety systems.

It is in the nature of humans to try to further our understanding of the world around us. No law will stop people undertaking research which might carry them into new domains, and there will always be a risk that a new technology will be misused. Therefore, it is important to understand the results of research and to explain the advantages and possible risks to everybody who is interested. There are of course great benefits to be gained from nuclear energy and ionising radiation, particularly in (fission) power plants and in nuclear medicine (both diagnosis and therapy). Also nuclear fusion, the energy source of the stars, may well solve all of mankind's energy problems in less than a century.

The use and abuse of radiation concerns, among others, physicists, engineers, lawyers and healthcare professionals, as well as the general public. Everyone should be able to judge on the application of radioactivity in various fields himself without referring to experts.

This book was originally published as a longer, more technical volume called Introduction to Radiation Protection. That book itself originated from a series of lectures that one of the authors (C.G.) gave over a period of more than 40 years. The text was first published in German by Vieweg and updated later in editions by Springer. It has also been translated into English and Japanese. A translation into Turkish is in preparation.

This book would not have been possible without the help of a large number of people. In particular, the help of Dr. Tilo Stroh and Dr. Ulrich Werthenbach was invaluable in the creation of the earlier versions. We thank Dr. Cornelius Grupen for injecting ideas for the book, particularly on the structure of the material. In addition, M.R. would like to thank his wife Clare for her help and encouragement, and C.G. wants to thank his wife Heidemarie for her continuous support and patience.

The aim of this more accessible book is to simplify the complicated physics and mathematics of the original version so that interested members of the public will be able to judge on possible dangers of ionising radiation. There is often an antipathy and distrust when ionising radiation is discussed. Occasionally this distrust is justified, especially when irresponsible discharges of radioactive waste into the environment are concerned. However, we are surrounded by radiation from many different sources, especially from the natural environment: the food that we eat and the air that we breathe are to a certain extent radioactive. This unavoidable radiation serves as a good starting point for comparison when discussing additional radiation from technical installations, and we can keep in mind that most of the additional radiation results from diagnostics and therapy in medicine.

We have tried to make the field of radioactivity accessible to the layman with illustrations and examples that hopefully appeal to general experience. The aim of this book is to improve the understanding of the basics of radioactivity and to assess the radiation risks in comparison to the risks that we are used to taking every day without any consideration.

Claus Grupen (Siegen, Germany)
and Mark Rodgers (Leeds, UK), October 2016

Contents

Chapter 1
Why Should I Read This Book?

Too many radiologists still believe there is a risk from a chest X-ray [exposure]. Few radiologists can explain radiation to the patient in words the patient can understand.

John Cameron 1922–2005

Life on Earth has developed under constant exposure to radiation. In addition to ionising radiation from natural sources, a multitude of exposures from artificial sources produced by mankind came into play in the twentieth century. These radioactive sources were produced by some impressive developments in science and technology, and in turn they helped spur further developments, including of course in medical diagnostics and therapy.

Humans have no senses for ionising radiation. Perhaps this is because radiation from the natural environment does not present a hazard, and so there is no necessity for a warning. However, possible risks related to elevated levels of ionising radiation have often been underestimated. For example, it has happened several times that sources have been disposed of illegally, and later found by children. Since there is no apparent danger from these sources (no smell or taste is noticed, and nothing untoward can be seen) they are sometimes handled by the children and even stored in their homes. Considering the strength of radioactive sources used in medicine and technology, the irradiation from these sources over a period of several days can easily lead to radiation sickness and even death.

In the fifties and sixties of the last century, people were enthusiastic about nuclear energy. The nuclear weapon tests in the atmosphere in the early sixties led to high levels of radioactivity in the air and on the ground. Almost nobody was interested in this pollution at the time. One of the authors of this book (C.G.) was involved in the sixties in measurements of the radioactive pollution in the air at Kiel University. The level was so high that the local radio station was asked to include the radiation level (in pico-Curies per cubic metre) in the hourly weather forecast, and they did do so for

© Springer International Publishing Switzerland 2016
C. Grupen and M. Rodgers, *Radioactivity and Radiation*,
DOI 10.1007/978-3-319-42330-2_1

Fig. 1.1. Portrait of Henri Antoine Becquerel (drawing by C.G.)

Fig. 1.2. Portrait of Wilhelm Conrad Röntgen (drawing by C.G.)

about six months. It was stopped after that, because very few people were interested in the radiation levels. People born in the sixties were actually 'radiating babies', because even breast milk was contaminated with radioactive strontium (strontium-90) and caesium (caesium-137). This view about radiation pollution changed drastically with the Chernobyl disaster, after which considerable concern was raised, even where the contamination was at a very low level.

To make appropriate judgments about the potential danger caused by radioactive sources, it is important to develop a feeling for the biological effects of ionising radiation. It is impossible to eliminate radiation exposure altogether: one cannot possibly avoid natural radioactivity from the environment. Therefore, additional exposures have to be assessed with reference to natural levels of radiation exposure. To estimate the potential risk from radiation from the environment and from other sources, a minimum knowledge about physics, chemistry and biology is required: this can effectively serve as a sense for radioactivity, which we would otherwise be missing.

Radioactivity was discovered by Henri Becquerel in 1896 (Figure 1.1), when he realised that radiation emerging from uranium ores could blacken photosensitive paper (paper that acts like a camera film). Originally it was believed that this was due to some fluorescence[1] radiation from the ore. However, the photosensitive paper was also blackened when the ore had not been previously exposed to light. The radiation spontaneously emerging from uranium was not visible to the human eye. Therefore, it was clear that he was dealing with a new phenomenon.

Shortly before this, in December 1895, Wilhelm Röntgen (Figure 1.2) discovered X rays. This radiation emerged from materials when they were struck by energetic electrons. Indeed, the discovery by Röntgen had been one factor stimulating Becquerel to investigate "fluorescence" radiation from uranium ore.

The new research field of radioactivity became particularly important when, in 1898, Marie and Pierre Curie succeeded in isolating new radioactive elements (polonium and radium) from uranium ore. Marie Curie was awarded two Nobel Prizes for her research (one in physics, one in chemistry).

At the turn of the century (1899–1902), investigations by Ernest Rutherford clearly demonstrated that there are different types of ionising radiation. Since initially it was impossible to identify these different types, they were named after the first three letters of the Greek alphabet: α ("alpha"), β ("beta") and γ rays ("gamma"). It was shown that α and β rays could be deflected by magnetic fields, and that γ rays could not (see Figure 1.3).

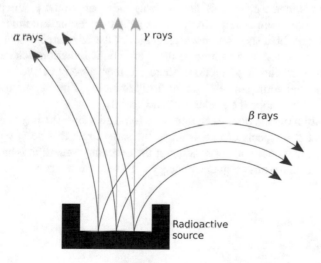

Fig. 1.3. Deflection of α, β, and γ rays in a magnetic field. The magnetic field direction is downwards, into the page

---

[1]The giving out of light of one frequency (colour) by a material after it has been exposed to light of other frequencies.

This radioactivity was at first an unalterable phenomenon of the natural environment. Nobody was able to turn inactive materials into radioactive sources by chemical techniques. Nuclear physics technology, however, continued to develop, and in 1938, Otto Hahn and Fritz Straßmann succeeded in inducing fission of uranium nuclei (although this discovery was accidental, as explained in Chapter 9).

The importance of radioactivity and of radiation protection for mankind and the environment is substantial. The judgment on the effects of ionising radiation on humans should not be left only to so-called experts. Everybody who is prepared to get involved in these problems should be in a position to come to his own judgment. It is highly desirable that, for example, discussions on the benefits and risks of nuclear power and disposal of radioactive waste are not dominated by emotional antipathy or blind support, but rather by solid facts about radiation and radiation-related effects.

So that the reader can get the most out of the discussions in the rest of the book, a few essentials of nuclear physics are needed, so these are provided in the next chapter. The book then goes on to explain the interactions of radiation in matter, and the ways radiation can be made and used by mankind (radiation sources and X rays). The next two chapters are on radioactivity in our (natural and man-made) environment, and the biological effects of radiation. There then follow chapters on the usage of radioactivity for good and ill, in power stations and weaponry, and one on radiation accidents. Because most of this book concerns ionising radiation, and there is currently public debate about other kinds of radiation, such as mobile phone signals, there is then a chapter on non-ionising radiation. Some practical aspects of radiation protection are covered in the subsequent chapter, before the main body of the book closes with a short summary chapter. For the interested reader, two further chapters are provided as appendices: these are about radiation detectors, and the organisation of radiation protection. There are then a copy of the periodic table, for reference, and some suggestions for further reading. Finally, there is a detailed glossary, which we hope the reader will find useful.

The intention of this book is to introduce the reader into the physical, technical, medical and legal aspects of radiation. At the same time, this book will hopefully contribute to the reader's understanding of the necessary scientific issues, allowing, for example, discussions on nuclear energy to be better informed.

Actually, I was expecting radio-activity to be something different . . .

Chapter 2
What Are Radioactivity and Radiation?

All composed things tend to decay.

Buddha 563–483 B.C.

We and everything around us are made of atoms. The idea of the atom came out of ancient Greek philosophy: some philosophers (notably Aristotle) thought that matter could be divided into ever smaller parts, and others (notably Democritus) thought that at some point, there was a minimum size of object, which they called "atomos", meaning "unsplittable". With the techniques of modern science, it has become clear that these atoms do exist, and each one is about a ten-billionth of a metre across: the full stop at the end of this sentence is about a million atoms across.

Ironically, we now know that atoms are not unsplittable: they are made up of smaller components. Each atom is composed of a small, positively charged nucleus, with some (negatively charged) electrons orbiting around it. The nucleus at the centre of the atom contains almost all of the atom's mass, but is a small fraction of the size, about a million-billionth of a metre across. This means that, in size terms, the nucleus inside an atom is in the same proportion as a grain of sand in a hot air balloon. A diagram of an example atom, with the size of the nucleus exaggerated, is shown in Figure 2.1.

Because very large and very small numbers occur frequently in this book, we will sometimes use the notation using powers of ten. This is a number written as a superscript by the number ten, and indicates how far the start of the number is from the decimal point.[1] Large numbers have a positive power of ten (so $1\,000\,000$ is written 10^6), and small numbers have negative power of ten (so $0.000\,001$ is written

---

[1]This can also be thought of as "how many times we should multiply (positive superscript number) or divide (negative superscript number) by ten". So 10^4 is $10 \times 10 \times 10 \times 10$, which is $10\,000$; 10^{-3} is $\frac{1}{10} \times \frac{1}{10} \times \frac{1}{10}$, which is 0.001; and 4×10^{-2} is $4 \div 10 \div 10$, which is 0.04.

© Springer International Publishing Switzerland 2016
C. Grupen and M. Rodgers, *Radioactivity and Radiation*,
DOI 10.1007/978-3-319-42330-2_2

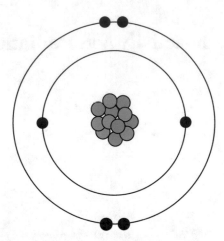

Fig. 2.1 A diagram of an atom of carbon-12. The size of the nucleus is greatly exaggerated so that its components are visible. There are six protons (pictured blue) and six neutrons (orange) in the nucleus. The positive charge of the nucleus is compensated by six electrons, arranged into two groups, called shells (two in the innermost shell and four in the next)

10^{-6}). We will also frequently use the power prefixes, shown in Table 2.1. Some of these will be familiar to most readers, perhaps in forms like kilograms (1 000 g), Gigabytes (1 000 000 000 bytes) and millimetres (0.001 m).[2]

Let us rewrite the sizes of the atom and the nucleus in these ways: an atom is typically 10^{-10} m or 0.1 nm across, and the nucleus at its centre is more like 10^{-15} m or 1 fm wide.

Table 2.1 Power prefixes for units

Prefix	Abbreviation	Power of ten	Number
Tera	T	10^{12}	1 000 000 000 000
Giga	G	10^{9}	1 000 000 000
Mega	M	10^{6}	1 000 000
kilo	k	10^{3}	1 000
milli	m	10^{-3}	0.001
micro	μ	10^{-6}	0.000 001
nano	n	10^{-9}	0.000 000 001
pico	p	10^{-12}	0.000 000 000 001
femto	f	10^{-15}	0.000 000 000 000 001

---

[2] A word of caution is in order here: in the USA and most other parts of the world, a billion is 10^{9} while in some parts, including Germany and other European countries, a billion is 10^{12}. This book will follow the US convention when writing numbers as words.

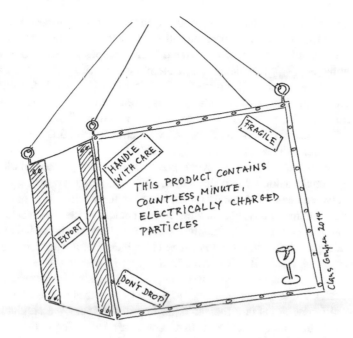

The nucleus and the surrounding electrons are charged objects. This charge ("electric charge", in full) means that they can interact with electromagnetism, and indeed the nucleus and the electrons are bound together by their mutual electromagnetic attraction. It is possible for many atoms to be joined together into a molecule: these are bound together by a residual electromagnetic interaction between the nuclei and the electrons of the different atoms.

The particles of the nucleus are called nucleons, and there are two types: positively charged protons, and neutral neutrons. Protons and neutrons have very similar masses: the neutron is 0.1 % heavier. The number of protons in a nucleus (the proton number) defines the chemical element: that is to say that the chemical behaviour of an atom is completely determined by the number of protons in its nucleus. For a neutral atom, there will be the same number of electrons in orbit as there are protons in the nucleus. If there are more or fewer electrons than there are protons, there will be a net charge: the atom is then called an ion.

The neutrons are essential in binding the nucleus together, and the number of them is called the neutron number. The *atomic mass* is the number of nucleons in that nucleus, and so is the sum of the proton and neutron numbers. It is possible for different nuclei to have the same proton number, i.e. to be the same element, but to have different numbers of neutrons. For example, all atoms of oxygen have 8 protons, but different atoms of oxygen might have 7, 8, 9 or 10 neutrons (oxygen atoms with 8 neutrons are by far the most common type). These different types of atoms, of the same element but with different numbers of neutrons, are called isotopes. Isotopes which are radioactive are called radioisotopes. Of the four isotopes of oxygen mentioned above, the first is a radioisotope, and the other three are stable (non-radioactive) isotopes.

In this book, two different notations are used to refer to isotopes. The first is a longer form, which is the element name followed by the atomic mass: for example carbon-12 or oxygen-18. In a more shorthand form, isotopes will often be referred to using the notation $_{\text{Proton number}}^{\text{Atomic mass}}$Element. The elements are written using their standard abbreviations: these are usually intuitive (like H for hydrogen and Al for aluminium) but some are not obvious, so a periodic table is given at the back of the book, listing all the elements with their abbreviations. Since the name of the element and the proton number are effectively the same piece of information, that number is often left out, so caesium-137 can be written $_{55}^{137}$Cs or ^{137}Cs.

Within any nucleus,[3] there are several positively charged protons in close proximity, so one might naturally expect them to fly apart, given the (electromagnetic) repulsion between them. Fortunately for us, there is a force which is about 100 times stronger, rather unimaginatively called the strong nuclear force, which acts to bind a nucleus together. In the same way that the electromagnetic force is carried by photons (particles of light), the strong force is carried by gluons, which are hard to observe directly as they are confined within a nucleus.

Protons and neutrons are themselves composite objects which consist of quarks (see Figure 2.2). These quarks are bound together by the strong nuclear force. Then, just as the different atoms in a molecule are bound together by a residual electromagnetic force, the different nucleons in a nucleus are bound together by a residual strong nuclear force.

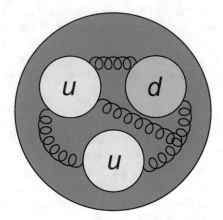

Fig. 2.2 A proton. Each proton contains three quarks (two of type "up" and one of type "down"), and multiple gluons, shown here as spring-shaped lines, gluing them together. The neutron has the same structure, but two of its quarks are of the "down" type and one is of the "up" type. Quarks have electric charge: "up" quarks have $+\frac{2}{3}$ of a unit and "down" quarks have $-\frac{1}{3}$

---

[3]With one exception: hydrogen nuclei, by definition, contain only one proton.

Well, that's one way to remember the word

2.1 Radioactivity

The binding of nucleons in a nucleus is stronger for some isotopes than others, i.e. some nuclei are sufficiently tightly-bound that they are stable, others have weaker binding, and would be in a more tightly-bound configuration if they changed their form: these are the unstable isotopes, which undergo radioactive decay. When we have a group of unstable nuclei of the same type, they will tend to decay, and it is useful to think of the half-life, which is the time taken for half of the nuclei to decay. This is explained in Section 2.2.

Figure 2.3 is a plot of proton number against neutron number. The combinations of numbers of protons and neutrons which make stable nuclei fall into a band called the *valley of stability*. Nuclei which have numbers of protons and neutrons which put them inside this band are likely to be stable, and moving further from the valley in either direction, the nuclei become ever more unstable (i.e. have shorter half-lives). The plot shows that, for stable nuclei of lighter elements, there are approximately as many protons as neutrons. For heavier elements, there are more neutrons than protons.

Protons, to the best of our knowledge, are stable.[4] Free neutrons, i.e. ones outside a nucleus, are not stable: they have a half-life of about 10 min. They decay into a proton and an electron (and an antineutrino, written $\bar{\nu}$, which is a very light, neutral particle[5]). This decay can be written as the equation:

$$n \rightarrow p + e^- + \bar{\nu}.$$

---

[4]Experiments are underway to find out if the proton is completely stable, or merely has an enormous half-life. The current limit says that the half-life of the proton is something over 10^{29} years, vastly longer than the age of the universe (13.8 billion years).

[5]There are in fact three types of neutrino, and three corresponding types of antineutrino, as explained in the glossary, but that is not important here.

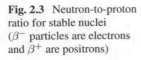

Fig. 2.3 Neutron-to-proton
ratio for stable nuclei
(β^- particles are electrons
and β^+ are positrons)

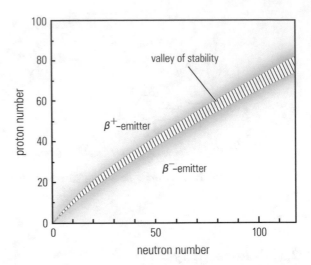

In a nucleus with excess neutrons (see Figure 2.3), the same process can occur for one or more of the neutrons, and in this case the proton stays in the nucleus, and the electron is emitted. This emitted electron is referred to as a β^-, pronounced "beta minus", and the nucleus as a β^--emitter.

Similarly, it is possible for nuclei with excess protons to be β^+-emitters. In this case, a proton decays into a neutron by emitting a positron (β^+), which is like an electron but positively charged, and a neutrino (this kind of decay is only possible inside an unstable nucleus: free protons cannot decay). However, there are two further possibilities for a nucleus with excess protons: α decay, which is discussed later in this section, and electron capture. In electron capture, one of the orbiting electrons is absorbed by one of the protons in the nucleus, leaving a neutron (and causing the emission of a neutrino).[6]

The difference between a β^- and a β^+, between electron and positron, is an important one in all of particle and nuclear physics. The electron and positron are each other's antiparticles: they are particles identical in all ways (such as their masses, stabilities and the absolute size of their charges) except they are of opposite charge. All particles have antiparticles, although for some, such as the photon, the antiparticle is the same as the particle.

The original (*parent*) nucleus and the resulting (*daughter*) nucleus have certain specific energies. This means that the β decay between them will give off a certain, specified amount of energy. However, because this energy is shared between the two emitted particles (electron and antineutrino, or positron and neutrino), there is a range of possible β particle energies, up to a certain maximum energy. In this maximum case, the β particle is taking all the energy, and the (anti)neutrino effectively none.

---

[6]This combination of a proton and electron into a neutron is exactly what happens when the dense core of a star is collapsing after a supernova explosion, and leads to the formation of a neutron star.

After a β decay, the daughter nucleus is often not in the most stable arrangement for that nucleus: it is in an *excited state*.[7] Excited states are marked in this book with an asterisk (*). These states move to the most stable (*ground*) state by emitting γ (pronounced "gamma") rays. These γ rays are photons (particles of electromagnetic radiation) of high energy. They are therefore physically the same phenomenon as light, but with dramatically higher frequency and tiny wavelength. It is normal for people to think of light as a wave, but it can equally be thought of as a particle, and these descriptions do not contradict each other. In this book, the particle description is the more useful one.

Since the energy difference between the excited nucleus and the ground state is fixed, these γ rays have a single, defined energy for each transition, rather than the continuous range described above for β particles.

When thinking about the energy of a single particle, or the average energy of a group of particles, it is useful to have a very small unit of energy, which is of roughly the scale of the energy carried by individual particles. The most common unit for this in particle and nuclear physics is a funny little unit called the electron-Volt (eV), whose origin is explained in Chapter 5. This unit, about 1.6×10^{-19} Joules, is so small that it would take several billion electron-Volts to lift up a grain of sand by one centimetre! In an old-fashioned TV set, each electron is accelerated up to an energy of about 20 keV.

As an example for β decays let us consider the decay of caesium-137:

$$^{137}_{55}\text{Cs} \rightarrow \, ^{137}_{56}\text{Ba}^* + e^- + \bar{\nu}$$
$$\hookrightarrow \, ^{137}_{56}\text{Ba} + \gamma \, .$$

---

[7]Imagine a pile of rocks in which one of the rocks in the middle was suddenly made smaller. The pile would probably want to resettle, as do nuclei after this kind of change.

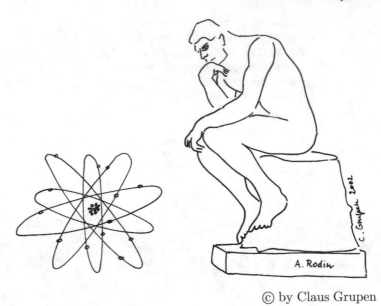

© by Claus Grupen

The thinker (about nuclear physics)

The ^{137}Cs gives off an electron of energy up to 514 keV, and leaves an excited nucleus of barium-137. That excited nucleus (^{137}Ba*) then decays, giving off a γ ray (which always has the energy 662 keV). In fact, it is also possible for this process to skip the excited state, and go straight to the barium's ground state, so no photon is emitted and there is more energy for the electron and antineutrino. This happens 5 % of the time.

The excited state of a nucleus will typically last for only a small fraction of a second (maybe 10^{-11} s) before decaying. Some nuclear excited states, however, last much longer. This excited state of barium-137 is an example: its half-life is two and a half minutes. These relatively long-lived nuclear excited states are called *metastable* states, and can be marked with a superscript "m", for example ^{137m}Ba.

Heavy nuclei, i.e. those containing a large number of nucleons, tend to decay by the emission of an α (pronounced "alpha") particle. This particle is composed of a pair of protons and a pair of neutrons, and is the same as the nucleus of a helium atom. This grouping of four nucleons happens to be a particularly stable and low-energy scenario, so this type of decay is usually preferred where it is possible, given the energies of the parent and daughter nucleus. This means that, where this decay and β^+ decay are both possible (which is frequently the case), it is usually α decay which takes place.

An example of an α-emitter is uranium-238, which gives out an α particle and transforms into excited states of thorium-234:

$$^{238}_{92}\text{U} \rightarrow \ ^{4}_{2}\alpha + \ ^{234}_{90}\text{Th}^* \ .$$

Since the nuclear levels have certain fixed energies, and only one particle is being given off in the decay, the emitted α particles from a particular isotope all have a single, characteristic energy, in the same way that γ rays from excited nuclei do.

One last possibility for the decay of heavy nuclei (for proton numbers over 90) is spontaneous fission. This is the process in which the nucleus splits into two lighter nuclei which are of broadly similar size, for example, californium-252 can fission into barium-148 and molybdenum-104:

$$^{252}_{98}\text{Cf} \rightarrow \, ^{148}_{56}\text{Ba} + \, ^{104}_{42}\text{Mo} \, .$$

In the earlier discussion on the valley of stability, the plot showed that heavier nuclei have proportionally more neutrons than lighter ones. This means that, after a fission, there will be an excess of neutrons, because the smaller nuclei created will still have proportionally as many neutrons as their larger parent. Usually, some of this imbalance is countered by the emission of free neutrons during the fission process (*prompt neutrons*), for example, four prompt neutrons are given out in the fission of fermium-256 to palladium-112 and xenon-140:

$$^{256}_{100}\text{Fm} \rightarrow \, ^{112}_{46}\text{Pd} + \, ^{140}_{54}\text{Xe} + 4n \, .$$

Frequently, there is still an imbalance in the fission products, so they will either undergo β decay, or give off neutrons directly (*delayed neutrons*), or both.

There are some isotopes which do not fission spontaneously, but can undergo *induced* fission if they are struck by one of these emitted neutrons. An example is the induced fission of $^{235}_{92}$U, given by the reaction

$$^{235}\text{U} + n \rightarrow \, ^{236}\text{U}^* \rightarrow \, ^{139}\text{I}^* + \, ^{96}\text{Y}^* + n \, .$$

There is one prompt neutron, and then (not shown in the reaction above) the highly excited iodine and yttrium nuclei will also emit a neutron each, and also undergo several β^- and γ decays to reach stable nuclear configurations.

In this example, a single free neutron was introduced, and then one prompt neutron and two delayed neutrons were emitted. The emitted neutrons can then go on and induce further fissions, and by continuing in this way, an increasing, or at least self-sustaining, chain reaction can be built up. This is the basis for both fission bombs and all currently operational nuclear power stations (see Chapters 8 and 9).

An excited nucleus left after a decay can release its energy in an indirect way: although normally it will release its excess energy by γ emission, it is also possible for this excitation energy to be transferred directly onto one of the electrons orbiting the nucleus: this electron will then leave the atom. These emitted electrons are called *conversion electrons*. However, the emissions need not stop there. The electron will leave a vacancy in its orbit (as will one taken in an electron capture), and one of the other orbiting electrons will fall in, releasing some energy. This energy can then be given out as an X ray, as shown in Figure 2.4.

Fig. 2.4 Illustration of the production of characteristic X rays. An electron has been removed from the atom (empty circle) and one from a higher shell moves down, emitting an X ray of characteristic energy (the energy difference between the shells)

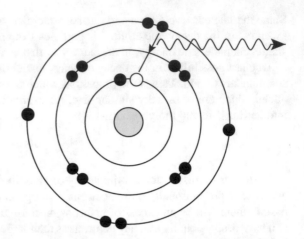

The most important properties of the different types of radiation mentioned in this chapter are compiled in Table 2.2. They are often referred to as *ionising radiation*, as they can produce ions from atoms when they interact, by removing electrons from them. This is explained in Section 3.1. It is worth noting that all these types of ionising radiation can be referred to as rays or as particles, and in this book, these terms are used interchangeably.

The X rays listed in the table deserve further explanation, as they are very relevant in the field of radiation protection. X rays are very short-wavelength electromagnetic radiation similar to γ radiation, but of slightly lower energy (longer wavelength). They are produced by the movements of charged objects: either during the transition of an electron between shells in an atom, or during the deceleration of a particle in an electric field (which can be within a material). X rays are discussed in more detail later in this book (see Chapter 5).

Table 2.2 Some properties of different types of ionising radiation

Type of radiation	Emitted particle	Typical energy
α ray	$^{4}_{2}He$	5 MeV
β^{-} ray	e^{-}	1 MeV
β^{+} ray	e^{+}	1 MeV
γ ray	γ	1 MeV
Neutron	n	1–6 MeV
X ray	γ	100 eV – 100 keV

2.2 Activity and Half-Life

The number of radioactive decays in a particular time is called the activity, and is measured in Becquerels (Bq). 1 Bq is one decay per second. The old unit Curie (Ci) is a very large unit, corresponding to the activity of 1 gram of radium-226:

$$1\,\text{Ci} = 3.7 \times 10^{10}\,\text{Bq} = 37\,\text{billion Bq}$$

$$1\,\text{Bq} = 27 \times 10^{-12}\,\text{Ci} = 27\,\text{pico-Curie}.$$

1 Bq is a very small activity. The human body typically has an activity of 9000 Bq (about 100 Bq per kg), which essentially originates from radioactive potassium (^{40}K) and radioactive carbon (^{14}C), which are predominantly ingested from normal food.

In radioactive decays, the number of nuclei decaying in a particular time is directly related to the number of existing nuclei. The number of radioactive nuclei decreases as the decay progresses, so the rate of decays will slow as time goes on. The decrease in the number of radioactive nuclei depends on the type of nucleus (isotope): each one has a half-life, which describes its rate of decay. The half-life indicates the time taken for half of the original nuclei to decay. Then after two half-lives, only one quarter of the original nuclei remain. Half-lives can vary between tiny fractions of a second and billions of years, depending on the specific isotope. Every half-life decreases the number of nuclei by a factor of two, as shown in Figure 2.5. Let us take an example from nuclear medicine: a patient is administered 0.16 nanograms of the radioactive

The fish on the right has the higher activity. Both fish would breach moderately cautious safety standards.

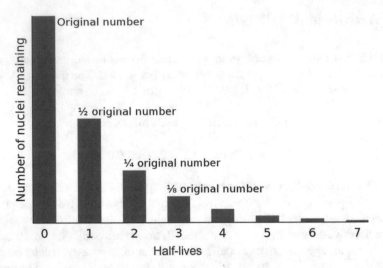

Fig. 2.5 The radioactive decay law

element iodine-131 (which has a half-life of 8 days) for the diagnosis of a possible thyroid cancer. Then after 32 days (4 half-lives), the amount of iodine-131 will have halved four times, i.e. one sixteenth of the original amount will remain, namely 0.01 ng. Only for very long times will the amount of iodine-131 approach zero.

Just as the number of radioactive nuclei decays in time, so does the activity. In our earlier example, the iodine-131 starts with an activity of 740 kBq, and after 32 days, the activity is about 46 kBq. Radioactive sources with a large half-life naturally have lower activities for a given number of nuclei. Frequently it is said that long-lived radioactive waste (e.g. with a half-life of a million years) is dangerous, because it has to be stored safely for this long period. This is not the whole story, because the long half-life means that the decay rate is rather low, so the immediate risk from these long-lived isotopes is relatively small.

2.3   Radiation Doses

Understanding the activity in Becquerels is the first step to finding out about any possible biological effects, but does not tell us a great deal by itself. This is because, apart from the absolute amount of energy absorbed in the body, the biological effects are related to:

the size of the exposed area or body – for the same absolute amount of radiation, a larger body will feel less effect, so the measures of radiation exposure must be inherently about the exposure per kilogram of tissue.

the radiation type – some types of radiation are intrinsically more damaging than others.

the distribution of the dose – some tissues of the body are more sensitive to radiation than others.

In this section, the units taking account of each of these factors in turn will be discussed. Starting with the amount of energy absorbed, the *energy dose* (measured in Grays (Gy)) is worked out by looking at the exposure per kilogram. Then the *equivalent dose* (measured in Sieverts (Sv)) is worked out by also taking account of the radiation type. Finally, the *effective dose* (also measured in Sv) is found by additionally taking account of the different tissues exposed. This last measure takes account of all the effects, and is the standard measure used in this book. Later in this book, when talking of a "dose", this effective dose is what is meant.

The energy dose is the amount of energy absorbed per unit mass. Because energy is measured in Joules, and mass is measured in kilograms, it is possible to talk of this quantity in Joules per kilogram. In this context, one Joule per kilogram is called one Gray (Gy). The old unit rad (radiation absorbed dose), still in use in the United States, is one hundredth of a Gray.

Apart from these units, another quantity can be used for the amount of charge created by radiation, the Roentgen (R). This rather specialised unit is sometimes still used in medicine. 1 Roentgen is equivalent to 8.8 mGy.

The Gray, the rad and the Roentgen describe the pure physical energy absorption. These units cannot easily be translated into the biological effect of radiation. As explained in Chapter 3, some radiation types have a more concentrated effect than others. Electrons, for example, ionise relatively weakly while, by contrast, α rays

give a high ionisation density, which means more concentrated damage. Therefore, the normal biological repair mechanisms are less likely to be effective after damage caused by α rays.

2.3.1 Weighting Factors

In order to take account of the biological effectiveness of each radiation type, the *radiation weighting factors* are used. The energy dose is multiplied by the relevant radiation weighting factor to find the equivalent dose for a particular radiation type. The equivalent dose is measured in Sievert (Sv). The old unit rem (roentgen equivalent man), still in use in the US, is one hundredth of a Sievert.

The radiation weighting factors depend on the type of radiation and for neutrons also on their energy. The definition of radiation weighting factors, following the recommendation of the International Commission on Radiological Protection (ICRP), is given in Table 2.3. The table also mentions muons: these are short-lived particles which are produced by cosmic radiation (see Chapter 6).

As an example, let us assume that a radiation worker in a reprocessing plant is accidentally exposed to an energy dose of 20 mGy from α particles, and one of 50 mGy from γ radiation. Then his equivalent dose is given by the sum of these energy doses with their weighting, i.e. by (20 mGy $\times$ 20 + 50 mGy $\times$ 1) mSv = 450 mSv. This is quite a significant dose, because the lethal dose of radiation for humans is 4000 mSv (the lethal dose is the one giving a death rate of 50 % within a month).

In many cases, particularly when dealing with exposures caused by specific events (such as the taking of an X-ray image), a person will have an exposure which is limited to one region of the body. In these cases, there is often a significant dose (remembering this is a per-kilogram measure) for one tissue, and a small or irrelevant dose for other parts of the body. In such cases, it is necessary to convert the partial-body dose into a whole-body dose, to understand the extent of the damage. For this, a tissue weighting factor has to be assigned to the irradiated organs of the body. This effective dose (also known as the effective dose equivalent) is obtained by multiplying the partial-bodydose received with the corresponding tissue

Table 2.3 Radiation weighting factors

Type of radiation	Factor
γ and X rays, all energies	1
Electrons and muons, all energies	1
Neutrons, energy under 10 keV	5
Neutrons, energy 10–100 keV	10
Neutrons, energy 100 keV–2 MeV	20
Neutrons, energy 2–20 MeV	10
Neutrons, energy over 20 MeV	5
Protons, all energies	2
α particles, fission fragments, heavy nuclei	20

Table 2.4 Tissue weighting factors

Organ or tissue	Factor
Red bone marrow	0.12
Colon	0.12
Lung	0.12
Stomach	0.12
Chest	0.12
Gonads	0.08
Bladder	0.04
Liver	0.04
Oesophagus	0.04
Thyroid gland	0.04
Periosteum (bone surface)	0.01
Skin	0.01
Brain	0.01
Salivary glands	0.01
Other organs or tissue	0.12

weighting factor for the organ in question. If more than one organ is concerned, one has to sum up the various contributions by multiplying the partial-body doses with the appropriate tissue weighting factors. The tissue weighting factors are given in Table 2.4.

For the purposes of radiation protection it is simply defined that the human body has fourteen 'organs' (and one further category for everything else). The tissue weighting factors are scaled so that their total is 1, and so they can be thought of as the percentage importance of that part of the body with respect to radiation damage. The thyroid typically weighs about 20 g, or 0.03 % of bodyweight, but its tissue weighting factor is 0.04, i.e. from a radiation damage point of view, it has an importance of 4 %, because it is so sensitive to radiation. In contrast, "other organs and tissue" is usually over 70 % of the bodyweight, but has only 12 % importance because it does not include any organs with high sensitivity to radiation.

The object of this calculation is that a non-uniform irradiation of the body with a particular effective dose should bear the same radiation risk as a homogeneous whole-body irradiation with an equivalent dose of the same size.

To give an example: a radiation worker has collected a broken radioactive source with his bare hands (skin dose 20 mSv), which he should not have done, and carried it in his lab coat pocket (5 mSv each for the lung and chest). The effective dose is then worked out to be $(20 \times 0.01 + 5 \times 0.12 + 5 \times 0.12) \, mSv = 1.4 \, mSv$.

"Atomic progress is quite educating!"

after Jupp Wolter

If a dose is restricted to a particular part of the body, then naturally, the effective whole-body dose that is calculated from it will have a smaller headline number. This can be important in understanding the scale of the dose. For example, let us imagine a partial-body dose of 4 Sv has been applied to the brain of a patient for cancer treatment, with no exposures for any other tissues. This looks initially like the procedure has administered a lethal dose, until we realise that the number given is not an effective whole-body dose, but a localised one. The effective dose here is 40 mSv (because the weighting factor for the brain is 0.01). This is still a significant dose, and would not be administered lightly, but it would not be lethal.

In principle, it is important to consider further radiation-relevant factors such as an increase in the risk of an additional dose after a high first dose, or reduced biological effects from an irradiation split into several separate doses.[8] This technique of intermittent radiation is used in cancer therapy: if a patient needs to receive, say, an effective dose of 2 Sv to destroy a tumour, this dose might be applied in ten separate fractions of 0.2 Sv, because in the intervals between these fractions, the healthy tissue will recover more easily than the cancerous tissue will. A typical interval between subsequent fractions of irradiation is a day.

To consider the time dependence of a dose, we use the *dose rate*: the change of the dose in a particular time interval. The dose rate may change rapidly, particularly for radioactive sources with short half-lives. A typical dose rate due to normal environmental radiation for humans is 0.3 μSv per hour.

---

[8]Historical attempts to do this have led to the measures Relative Biological Effectiveness (RBE) and Quality factor, which we do not consider further here.

2.3.2 Avoiding Doses

It makes sense intuitively that being more distant from a source of radiation reduces the dose received from it. Indeed, just from the spreading out of the radiation, for most sources, the intensity falls with the square of distance (the exception being when the source is directed, as it is in an X-ray tube). This means that being twice as far away from a source reduces the dose by a factor of four. In addition, the intensities of α and β rays are reduced significantly even by a few centimetres of air, due to absorption (see Chapter 3). This means that the dose received falls markedly with increasing distance from the source.

To avoid unnecessary doses, it is important to remember some simple rules for the handling of radioactive sources:

- keep the activity of the source as low as reasonable,
- keep a sufficient distance away from the radioactive material, if possible,
- keep the exposure time as short as feasible, and
- use shielding whenever possible.

Every person, whereever he or she is in the world, is exposed to natural radioactivity from cosmic radiation, terrestrial radiation, incorporation (ingestion (eating and drinking) and inhalation (breathing)), and the activity of his or her own body. This number averages about 2.5 mSv per year, but varies a little depending on where the person lives. Higher values occur in areas where radioactive elements occur in the soil (e.g. radioactive thorium). Additional radiation, for example due to nuclear power plants or other technical installations, is limited by most national regulations to 1 mSv per year. This chapter ends with an extra section about putting this kind of dose into context.

Summary

Ionising radiation is released in most nuclear transformations. α rays are helium nuclei. β sources emit high-energy electrons (β^-) or their antiparticles (β^+, called positrons). α and β decays alter the chemical nature of an element. γ rays, which are high-energy photons, are often emitted by a nucleus after α or β decay. Fission (spontaneous or induced) is also possible, and the neutron excess in fission products can be reduced by prompt or delayed neutron emission, or by β^- decays. The essential units of radiation protection are the Becquerel (Bq) for the activity and the Sievert (Sv) for the effective dose (which includes weightings for the biological effectiveness). This is fundamentally a per-kilogram measure, because a larger person will be able to absorb a larger absolute amount of radiation for the same amount of damage. Each radioactive isotope has a particular half-life. Radioisotopes with large half-lives are associated with a low activity, and those with short half-lives with a high one. The dose from a source can be reduced by keeping a distance from it, reducing the exposure time, and using shielding.

Putting doses in context

The common units for radiation doses, such as Sv, can be hard for the layman to interpret. Even low radiation exposures, which can only be detected by extraordinarily sensitive measurement devices, occasionally lead to over-reactions in public discussions. It is very important to express radiation levels, caused, for example, by the transport of nuclear waste (see Chapter 12), or by nuclear power plants, in units which can be understood and interpreted by the layman. What the public needs is a good sense of scale, so that the general size of particular radioactive exposures can be understood in context.

Mankind has evolved and developed alongside perpetual natural radiation, and there are no hints whatsoever that this radiation has created any problems for our biology. The natural radiation dose is subject to regional variations, but the natural annual whole-body radiation dose does not fall below 2 mSv for anybody (it is several times higher than this in some places). This natural annual dose sets the scale on which to judge on additional man-made radiation burdens, for example, during medical diagnosis.

This 2 mSv annual minimum should provide the reader with a useful yardstick for the scale of doses discussed in this book and elsewhere. It should be a sufficiently simple and memorable number, so that anyone of any background can make a realistic, informed judgment about radiation risks. Some typical examples are provided below.

Type of radiation exposure	Dose in mSv	Years of minimal background dose
Eating one normal banana	0.0001	0.00005
Dental X-ray image	0.01	0.005
Annual exposure to the public from nuclear power plants	under 0.02	under 0.01
Worker transporting radioactive waste	under 0.03	under 0.015
Flight from London to New York	0.03	0.015
X-ray image of the chest	0.03	0.015
Mammography	0.5	0.25
Scintigraphy of the thyroid gland	0.8	0.4
Annual dose for heavy smoker	1	0.5
Positron-emission tomography (PET)	8	4
Computed tomography (CT) of the chest	10	5
Annual limit for exposed workers in Europe	20	10
in the USA	50	25
Maximum lifetime dose for exposed workers in Europe	400	200
Lethal dose	4000	2000

Chapter 3
What Does Radiation Do?

A careful analysis of the process of observation in atomic physics has shown that the subatomic particles have no meaning as isolated entities, but can only be understood as interconnections between the preparation of an experiment and the subsequent measurement.

Erwin Schrödinger 1887–1961

Strictly speaking, radiation is never directly measured: it can only be detected via its interaction with matter. There are multiple different ways in which radiation can interact with matter: these interactions are characteristic for each radiation type – charged particles, neutrons, and photons (X and γ rays).[1] Similarly, there are different ways of detecting these different radiation types, which are covered in detail in Appendix A.

3.1 Interactions of Charged Particles

In passing through matter, a charged particle will frequently strike an electron within an atom, and pass some energy to it. There are then two possibilities. If the electron has been given enough energy, it can escape the atom entirely. Because a (negatively-charged) electron has left a (neutral) atom, what is left is positively charged, and is called an ion, and so this process is called ionisation. The second possibility is that the struck electron stays in the atom, but moves to a higher-energy position, further from the nucleus, called an excited state. These excited states are marked with an asterisk (*). An excited atom will decay some time (typically a nanosecond) later: i.e. the electron will fall back into the lower energy state, and a photon will be given out. For example, for α particles, an interaction means either

---

[1]Neutrinos, although they are a type of radiation, interact so weakly with matter that they are irrelevant in terms of radiation doses, so they are not discussed in this chapter.

© Springer International Publishing Switzerland 2016
C. Grupen and M. Rodgers, *Radioactivity and Radiation*,
DOI 10.1007/978-3-319-42330-2_3

$\alpha + atom \rightarrow atom^* + \alpha$, then: $atom^* \rightarrow atom + photon$ (excitation),

or

$$\alpha + atom \rightarrow atom^+ + e^- + \alpha \qquad \text{(ionisation)}.$$

The rate of energy loss of a charged particle passing through matter depends on the particle's speed, on its charge, and on the properties of the material it is passing through. A slow particle spends a relatively large amount of time close to individual atoms: correspondingly, the probability that it will interact with any specific atom it passes is relatively large. The particles interact via the electromagnetic force, and so when the moving charged particle has a stronger (electromagnetic) charge, it will interact more frequently, and give up more energy in each collision, so it will lose energy more quickly (similarly, if the particles being struck were more strongly charged, this would lead to faster energy loss, but the struck particle is usually an electron, and all electrons have the same charge). Also, in general, more interactions will happen, and energy loss will be faster, in a denser material.

Although this varies according to the specific isotope, a particle from a radioactive source will typically have an energy of a few Mega-electron-Volts. The average energy loss of electrons[2] of these energies in air is approximately 0.25 keV per millimetre. This means they have a range of under 10 m in air, or 0.3 mm in lead. α particles lose energy much more quickly (about 100 keV per millimetre in air) because of their stronger charge and higher mass. The stronger charge (twice the size of the electron's) increases both the probability of interactions, and the energy transferred in each interaction. The higher mass means that at similar energies, the α particle is moving much more slowly, increasing the interaction probability further. This means that α rays are rather short-ranged: just 4 cm in air, or 25 μm in aluminium (approximately). The great benefit of this short range is that it is almost never necessary to shield against α particles produced outside the body: any clothing or a small distance in air will reduce their impact to a negligible level. The corollary of this is that when α particles are produced inside the body (for example by inhaled radon gas) they deliver stronger and more concentrated damage than an equivalent β ray would. The α particle gives up all its energy in a small region of tissue, causing concentrated damage. It is for this reason that the radiation weighting factor (see Table 2.3 in Chapter 2) for α particles is very high.

The fact that particles which move more quickly deposit energy more slowly has a clinically useful consequence. When protons or nuclei are accelerated and fired at the human body, the damage is concentrated at a particular depth, allowing a tumor to be targeted, as explained below.

As a fast-moving particle moves through the body, it will decelerate as it gives up energy to ionisation electrons in the tissue. At first, this deceleration will be gentle, as the particle is fast-moving. As it decelerates, however, the rate of energy loss increases (because slower-moving particles have a higher rate of energy loss). This means that a large fraction of the energy deposited will be at the very end of its track.

---

[2]The reader will remember that β^- particles are electrons.

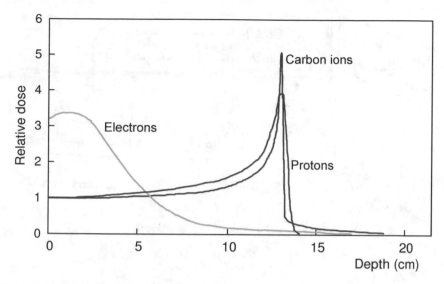

Fig. 3.1 A plot showing the energy loss (and therefore the dose delivered) at various depths in tissue, for 135 MeV protons (blue line), 3000 MeV carbon ions (red line) and 20 MeV electrons (green line). The Bragg peak is clearly visible for the protons and the carbon ions: almost all the dose is delivered in a very small range of depths. The carbon ions have a sharper delivery peak, but also have a slightly higher exposure at deeper positions, due to the occasional breakup of the carbon nuclei ("fragmentation")

This spike in the energy loss is called the Bragg peak. It is particularly pronounced when larger atomic nuclei (such as carbon or nitrogen) are used, as they have higher charge, and their larger mass means that they will be moving more slowly for a particular energy.[3] High-energy protons also show a strong Bragg peak. There are no radioactive processes which produce protons or heavy ions at high energies, so they must be produced in accelerators. Figure 3.1 shows the clear Bragg peak for 135 MeV protons and the sharper Bragg peak for 3000 MeV carbon ions. For comparison, an equivalent line is given for 20 MeV electrons. This does not show a similar sharp peak.

The Bragg peak is useful in non-invasive cancer treatment. It means that, when heavy charged particles are used (protons, or, better, heavier nuclei), a large part of the energy deposition occurs at a well-defined distance into the body. This means that the depth at which most of the radiation damage occurs is very predictable, and this depth can be varied by varying the energy of the particle beam. The horizontal position of the beam can of course also be adjusted. Using this control, it is possible to destroy very precisely regions of cancerous tissue (even deep within the body), while delivering relatively little radiation damage to the surrounding tissue.

---

[3]For light particles such as electrons, the Bragg peak effect is not visible because by the time the particle has slowed down enough for the interaction probability to rise strongly, it does not have much energy left.

This is called the raster-scan technique. Proton therapy and heavy-ion therapy are often referred to together, under the collective term *hadron therapy*.

So far in this section, only ionisation and excitation processes have been discussed, but there is one further interaction type which is common in particle interactions with matter, called bremsstrahlung.[4] In this case, there are not discrete events of specific electrons in the matter being struck by an incident particle: instead, the electric fields within the material (between the nuclei and electrons) cause the particle to decelerate, and to radiate energy in the process. The energy radiated comes in the form of photons, which are often in the X-ray range. The power output from this process has a strong dependence on the energy per unit mass of the particle, which means that although in principle both low-energy particles and heavy particles do experience this process, it only happens to a significant extent for high-energy, light particles. This means that the only practical importance of this mechanism comes in the context of high-energy electrons: for comparison, a proton of a given energy will give out by bremsstrahlung less than a millionth of the energy given out by an electron of the same energy. For an electron moving through a dense material (such as lead), the energy loss by bremsstrahlung is greater than that by ionisation and excitation if the particle's energy is over $10\,\mathrm{MeV}$ (approximately). Bremsstrahlung is important not only because it means that materials have a stronger stopping power for high-energy electrons than they might otherwise, but also because the X rays produced can have their own effects. It is (predominantly) by bremsstrahlung that X-ray tubes produce their X rays.

---

[4] A German word meaning "breaking radiation".

3.2 Interactions of Neutrons

Neutrons are neutral particles, meaning that they do not take part in electromagnetic interactions. This gives their behaviour in matter a very different character from that of charged particles. Neutrons (practically) never cause ionisation by striking the electrons within a material. They do, however, have interactions with matter that they are moving through. They are nuclear particles, and can interact (using the strong nuclear force) with other nuclear particles.[5] If a neutron strikes a nucleus, it can interact with it either simply by an exchange of momentum (and therefore energy), or by entering the nucleus and changing its character (which may lead to the emission of a different particle, such as an α particle or a proton). Some examples of nucleus-changing interactions which are used to detect neutrons are:

$$n + {}^{6}_{3}\text{Li} \rightarrow \alpha + {}^{3}_{1}\text{H},$$
$$n + {}^{10}_{5}\text{B} \rightarrow \alpha + {}^{7}_{3}\text{Li},$$
$$n + {}^{3}_{2}\text{He} \rightarrow p + {}^{3}_{1}\text{H},$$

and in addition, the interaction

$$n + p \rightarrow n + p,$$

which is only a transfer of momentum and energy, is also used. It is the moving charged particles which can then be detected, in the same way as if they were the initial incoming particles.

It was noted in the previous section that more slowly-moving charged particles interact more strongly with the material they are traversing. The same is also true for neutrons: when they move more slowly, they are more likely to interact with nuclei, particularly for the interactions which change the type of the nucleus. This means that higher-energy neutrons must first be decelerated ('moderated') before they can be detected efficiently using a nucleus-changing interaction. The most efficient moderation interaction is the momentum transfer between an incident neutron and a proton (i.e. a nucleus consisting entirely of a single proton). This means that, for moderators, materials containing lots of hydrogen in their molecules are a good choice, because a hydrogen nucleus is a single proton. Paraffin and water are common choices as moderators.

Because the neutron interaction can change the character of a nucleus, neutrons can *activate* a material they pass through, i.e. cause it to become radioactive. For example, natural cobalt is almost entirely composed of cobalt-59, which is stable,

---

[5]Protons, of course, also have the possibility to undergo these interactions. However, the nuclear interactions of protons are usually not significant, because the interactions with atomic electrons are much more frequent. Protons' nuclear interactions are particularly unlikely because both they and nuclei have positive charges, meaning they repel each other and so are less likely to meet.

and this can be activated by being transformed into cobalt-60 (a β^--emitter with a half-life of 5 years):

$$^{59}\text{Co} + n \rightarrow {}^{60}\text{Co} \, .$$

Different isotopes have different probabilities of being activated. There are some elements which have extremely low probabilities of neutron activation (such as helium and iron). This means that in the design of a system which will receive a large amount of neutron irradiation, such as the reactor vessel in a nuclear power plant, it is possible to choose materials which will suffer very little activation.

Neutrons are the only radiation type that cause a significant amount of activation. Although the activation of nuclei by α particles has been observed, it is very rare. β and γ radiation can in principle activate materials, but the probability is so low that the process can be completely ignored.

3.3   Interactions of Photons

Like neutrons, photons are not charged themselves, but will produce moving charged particles in interactions. The interactions of photons are fundamentally different from those of the other particles in this chapter because photons can be created (emitted) and destroyed (absorbed) freely. For other particles it makes sense to talk of the range (for each energy) they will have in a particular material. Photons, by contrast, have a potentially infinite range, but with an ever-decreasing intensity (and clearly, at some point, the intensity is so low as to be irrelevant).

The process of decreasing intensity of a beam of photons is called attenuation, and is due to a combination of three processes: the photoelectric effect, Compton scattering, and pair production.

The photoelectric effect is a type of ionisation – it is the liberation of an electron from the atomic shell, after it is struck by a photon:

$$\gamma + atom \rightarrow atom^+ + e^-.$$

In this process, the photon is completely absorbed, and a fast-moving electron is produced. Three criteria have an effect on the probability of this interaction. Firstly, it is more likely if the atom has a high proton number, simply because larger atoms have more electrons that might be hit. Secondly, there is a general decrease in the probability of this interaction as the photon energy increases. This is essentially because higher energy means shorter wavelength, and shorter wavelength photons are effectively smaller, and so less likely to strike an electron. Thirdly, if the photon's energy happens to match very closely the energy that an electron needs to escape the atom, the probability of this interaction is enhanced.

The photoelectric effect is the dominant one at relatively low photon energies (e.g. in the X-ray regime) and for heavy absorbers (such as lead and tungsten).

The Compton effect is another type of ionisation, in which the photon collides with an electron in an atom, but is not absorbed: it is purely a transfer of some energy (and momentum) from the photon to the electron. This is comparable to one ball hitting another in the game of pool: the photon can be thought of as the cue ball (white), and the electron as a colour ball. In equation form:

$$\gamma + atom \rightarrow \gamma + atom^+ + e^-.$$

The photon leaves with less energy than it had before. As for the photoelectric effect, the scattering probability is related to the number of potential scattering electrons in the atom, i.e. to the proton number. Unlike the photoelectric effect, which requires an almost head-on collision, the probability of this interaction does not fall (as strongly) with energy, as glancing collisions are possible. Because the probability of Compton scattering falls less strongly with increasing energy, this process tends to have greater impact on the attenuation of photon beams at higher energies.

After the Compton effect, and also after the photoelectric effect, an electron is missing in the atomic shell. There may be emission of one or more X rays from the atom as the vacancy is filled by an electron falling from a higher shell (and then the original place of this falling electron may be taken by one from a yet higher shell, and so on).

Pair production is the conversion of a photon into an electron-positron pair within the material,

$$\gamma + nucleus \rightarrow nucleus + e^+ + e^-.$$

This process is only possible inside a material, and that is because it relies on the presence of an electric field from a nucleus. Indeed, during this process, some momentum is transferred to the neighbouring nucleus. Clearly, this is another process in which the photon itself ceases to exist, and in this one, a certain amount of matter is created from energy. The energy required to create mass follows the famous equation $E = mc^2$. Because there needs to be enough energy to create the electron and positron, and also because some energy is transferred to the neighbouring nucleus, there is a minimum energy for this interaction. If the original photon has an energy far above this limit, pair production is the dominant interaction process.

The fall in intensity of a photon beam in a material due to the three processes described above can be described by a quantity called the mass attenuation coefficient. Figure 3.2 shows the mass attenuation coefficient for different photon energies, when lead is the absorber. The photoelectric effect dominates at comparatively low energies and also shows several enhancements where it matches certain electron transition energies precisely. Compton scattering is effective for intermediate energies, and then pair production is the main process at high energies. At very high energies, well above 10 MeV, photons will initiate electromagnetic cascades, where alternating processes of pair production and bremsstrahlung of electrons and positrons produce large numbers of photons and free electrons.

Fig. 3.2 The mass attenuation coefficient for photons in lead, for different energies

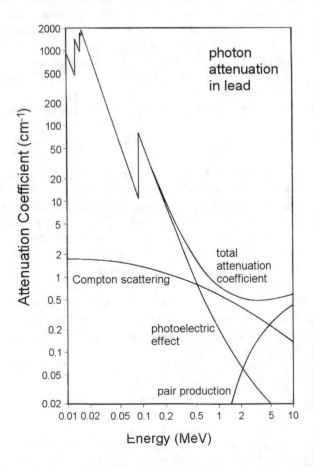

Summary

Charged particles lose their energy in matter mainly by ionisation and excitation. For electrons, there is an additional energy loss from bremsstrahlung. Because of the relatively large energy loss of charged particles, their range in matter is relatively short in most cases. In contrast to this, photons are attenuated only weakly when passing through matter, especially for the energies in the MeV range which are typical when discussing radioactivity. External radiation, therefore, usually consists of γ rays. Special care must be taken with neutrons, which have a relatively large range because they are electrically neutral. From the point of view of radiation protection, neutrons are very unpleasant because by hitting the nuclei of cells, they can create substantial radiation damage. They often produce charged particles in these interactions, which is how they can be detected. They can also cause some materials to become radioactive themselves (activation).

Chapter 4
How Can We Make Radiation?

> *Radioactive sources are used throughout the world for a wide*
> *variety of peaceful and productive purposes in industry,*
> *medicine, research and education, and in military applications.*
> International Atomic Energy Agency

One might assume that radioisotopes are the only significant sources of radiation. Instead, thanks to rapid developments in both fundamental physics research and its technical applications, there is a variety of possibilities for producing nearly all sufficiently long-lived nuclei, elementary particles and photons in the form of radiation sources. These sources span a huge range of energy, from ultracold particles (25 meV) up to energies of 1 TeV. If, in addition, cosmic rays are considered, particles with energies even in excess of 1 TeV are available, albeit with very low intensity. In the following sections, the main methods of production of ionising radiation are described, along with their important applications in medicine and in batteries.

4.1 Charged Particle Sources

All charged particles can be accelerated up to very high energies in accelerators, and then potentially stored within them. Most accelerators used in nuclear medicine are linear accelerators in the range below 100 MeV (see Figure 4.1).

In a linear accelerator (often abbreviated to "LINAC"), the acceleration itself takes place in a so-called cavity. This is composed of multiple rings around the path of the particle, whose charge is changed at a specifically chosen frequency. To illustrate how this works, consider a negative particle, perhaps an electron, moving along the beam line of Figure 4.1 (after the focussing, collimator and pre-buncher). It is on a path through the middle of the rings. The first ring is positively charged (to begin with), and the second starts negatively charged. The negative particle is attracted (and accelerated) towards and through the first (positive) ring, but then as

© Springer International Publishing Switzerland 2016
C. Grupen and M. Rodgers, *Radioactivity and Radiation*,
DOI 10.1007/978-3-319-42330-2_4

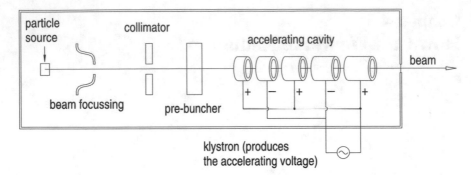

Fig. 4.1 A linear accelerator. The first three stages (focussing, collimator and pre-buncher) help keep the particles moving in a narrow beam. The cavity, composed of multiple rings, then accelerates the particles

it leaves the first ring and is moving towards the second, the charge flips. Now the first ring is negatively charged, and repels the particle onward, and the second ring is positively charged, and attracts it forwards. In this way, the particle can be accelerated continuously using many rings. The same story works for positive particles entering at a different time in the cycle.

In order to move particles to higher energies, it is necessary to run the particles through the cavities a large number of times, and the only practical way to do this is by using a circular accelerator called a synchrotron. In these rings, bending magnets are used to keep the particles in a narrow circular tube. Because it is important that there are no air molecules obstructing the accelerating particles, a vacuum is maintained in this tube. More complicated arrangements of magnets (quadrupoles) are used to keep the beam of particles focussed. Figure 4.2 is a sketch of the standard equipment of a circular accelerator. The name synchrotron comes from the fact that the strength of the magnetic guiding field has to be increased in a synchronous fashion with the increasing momentum of the particles: when the particles are moving more quickly, it takes a stronger field to cause their paths to bend to the same extent, i.e. to keep the particles on a fixed orbit in the vacuum pipe. The strength of the guiding magnets is increased by carefully increasing the current delivered to them.

Particles have to be injected into the circular accelerator at a particular point, and usually a linear accelerator is used to do this, so that the particles can start at a relatively high energy before entering the synchrotron. The particles most often used in these accelerators are electrons, positrons, protons and antiprotons. Typically, e^+ and e^- are accelerated to, say, 100 GeV in electron synchrotrons and p and $\bar{p}$ to a maximum of about 10 TeV in proton synchrotrons. If these high-energy particles are struck against a target, large numbers of new particles (then called secondary particles) will be created. These secondary particles can then themselves be turned into a focussed beam, and even accelerated around a synchrotron. For the purposes of research into high energy physics, beams of various particle types (pion, kaon, muon and neutrino) are produced by this technique. Beams of negative pions have

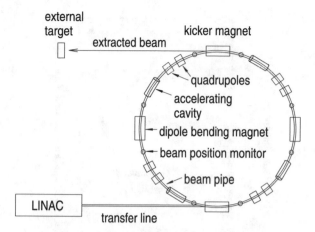

Fig. 4.2   A synchrotron. The bending magnets keep the particles running on the circular course. The strength of these magnets has to be increased over time, as the particles in the beam are accelerated

also been used in the field of medicine for tumour treatment. Heavy ions, which can also be accelerated in synchrotrons up to high energies, are also an excellent tool for tumour treatment in the context of hadron therapy (although hadron therapy can also use protons, as discussed in Section 3.1).

In electron synchrotrons at high energies, a substantial fraction of the beam energy is lost by synchrotron radiation (the emission of photons by the electrons as their paths are curved by the bending magnets). The production of synchrotron light at circular electron accelerators is certainly a disadvantage for particle physicists, as for them, it represents an energy loss. On the other hand, it can serve as a brilliant source of photons with energies up to the X-ray range. Using specific magnetic structures (wiggler magnets and undulators), highly intense beams of X rays can be generated, and they can then be used, for example, for structure analysis in solid state physics or biology, or for medical applications. Multiple facilities for producing and using this synchrotron light have been constructed, for example the European Synchrotron Radiation Facility in Grenoble (France) and the German Electron Synchrotron in Hamburg (Germany).

4.2   Photon Sources

X-ray tubes are a relatively old technology, but are still an effective source of moderately high-energy photons. They work by accelerating electrons between two charged plates, and letting them strike the positively charged plate (the anode). The electrons give off X rays as they decelerate within the anode (bremsstrahlung). In typical X-ray tubes, photons with energies up to several hundred keV or even a few MeV can be produced. X-ray tubes are discussed in more detail in Chapter 5.

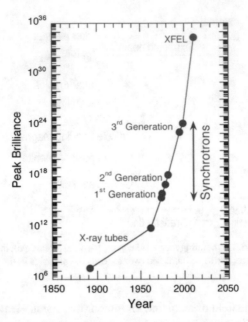

Fig. 4.3 The evolution of the brightness of X-ray sources over the last 120 years, starting from the low-brilliance X-ray tubes, then through various generations of electron synchrotrons up to the X-ray free electron lasers. (The brilliance is a measure of the intensity of light of any given frequency, and has the rather complex unit of photons/s/mm^2/mrad^2/ 0.1 % bandwidth. Credit: www.psi.ch/swissfel/why-swissfel)

The photon energy range which is classically covered by X-ray tubes, is also obtained from synchrotron-radiation sources. However, the photon intensity of synchrotron-radiation sources is many orders of magnitude higher (see Figure 4.3).

The photons coming out as synchrotron radiation have a continuous range of frequencies (or equivalently of energies or of wavelengths). It is possible to select out only one specific frequency, using a device called a monochromator. For X rays, this consists of a crystal which makes particular frequencies move in particular directions. Because the original synchrotron radiation was so intense, even if only one of the frequencies is taken, there is still sufficient intensity in the beam. An example usage of monochromatic X rays from electron synchrotrons is the medical diagnostic technique of coronary angiography. In this non-invasive investigation, the coronary arteries are marked with stable iodine as a contrast agent. Two digital images are taken, one at a frequency strongly absorbed by iodine, and one which is barely absorbed at all by iodine. By inspecting the differences between these images (worked out by a computer programme), the medical professional can see an image only of the blood vessels, with the surrounding tissue not being visible. This is called the 'dual-energy technique' or 'K-edge subtraction technique'.

If a single output frequency is required, the highest X-ray intensities available come from so-called free electron lasers (FELs). These receive electrons from an accelerator and take them along a straight path, subjecting them to a very strong, varying magnetic field along the path. The electrons are pushed from side to side, to a small degree, along their path, and emit X rays in the process. Because FELs produce all their radiation in a narrow band of frequency, their "peak brilliance" (the output for their most intense frequency) is very high, although they cannot achieve the extremely high levels of overall X-ray output available from synchrotrons.

There is a fourth photon source which is useful in medical imaging, although it is rather indirect. It uses β^+-emitters in a technique called positron-emission tomography (PET). In this technique, normal bio-molecules, often glucose, are marked with a positron-emitting radioactive isotope (e.g. ^{18}F) and injected into the patient. The tagged molecules migrate to locations where the sugar is needed, e.g. into the brain. The positrons emitted from the radioisotope have a very short range (typically a few millimetres): each one annihilates with an electron from an ordinary atom, producing a pair of photons travelling in opposite directions along the same line ($e^+ + e^- \rightarrow \gamma + \gamma$). These photons are recorded in detectors which determine where the photons started, and therefore where the glucose is. In this way, PET shows the amount of glucose moving to each region of the body, and so indicates which cells are most active (using the most glucose) when the scan is taking place. This information can then be interpreted by doctors to diagnose problems with particular organs.

For completeness, the use of decays of radioisotopes as photon sources must be mentioned. After α or β activity, the daughter nuclei frequently decay by γ emission into their ground state. In these decays, photons with energies up to several MeV can be produced. A medical application of this is given in Section 4.5.

4.3 Neutron Sources

Because neutrons normally sit inside the nucleus, the production of free neutrons involves an interaction of a nucleus with another particle. In dedicated neutron generators, single neutrons can be produced by striking specific targets with (high energy) protons or with α particles. This typically provides neutrons with energies of a few MeV. Neutrons can also be produced in interactions of a photon with a nucleus, for example a deuterium nucleus (which is $^2_1$H, and is written d) can be decomposed:

$$\gamma + d \rightarrow p + n \,.$$

A classical technique is neutron production in reactions involving an α particle. In this technique, α-emitting radioisotopes are mixed with a beryllium isotope. These α rays interact with ^{9}Be to produce neutrons of around 5 MeV according to the reaction

$$\alpha + {}^9_4\text{Be} \rightarrow {}^{12}_6\text{C} + n \,.$$

As the α-emitter, one can use radium (^{226}Ra), americium (^{241}Am), plutonium (^{239}Pu), polonium (^{210}Po) or curium (^{242}Cm or ^{244}Cm). Although these processes produce only about one neutron for every ten thousand α particles, this remains a useful source because of the well-known energy of the neutrons.

In fission reactors, highly radioactive fission products are generated. Since the fission materials, e.g. $^{235}_{92}$U, are relatively neutron rich, the fission products contain too many neutrons. The neutron excess can be decreased by the emission of prompt or delayed neutrons (these are discussed in Chapter 2). Nuclear-fission reactors are therefore a rich source of neutrons. The fission products can also try to reach a stable final state by successive β^- decays.

In the Sun, which is a fusion reactor, four protons are fused into helium in a multi-stage process involving other nuclei. In future fusion power plants, deuterium (written d or ^{2}H) and tritium (written t or ^{3}H) will be fused to make helium:

$$d + t \ \rightarrow \ ^4\text{He} + n \,.$$

In this reaction, an energetic neutron is generated (with an energy of 14.1 MeV). The original hope, that fusion power plants would be completely clean from the point of view of radiation protection, is not fully tenable. These energetic neutrons are very difficult to shield against. They will activate the reactor materials and produce some quantity of radioisotopes, so that a fusion reactor also represents a potential danger from the point of view of radiation protection. On the other hand, it is desirable that energetic neutrons are produced, despite their disadvantageous effects on the materials, as they are the source of power for fusion reactors. As described in Section 8.2, the big advantage of fusion reactors is that uncontrolled chain reactions cannot occur, which is a well-known possible problem with nuclear-fission reactors.

In spallation neutron sources (SNSs), energetic protons produce a large number of neutrons in reactions with heavy nuclei: it is possible to create up to 30 neutrons per reaction in this way. SNSs provide highly intense neutron beams for scientific research and industrial development. The high intensity of beams from SNSs means they can be used to make very precise measurements. In addition, complex samples from physics, chemistry, materials science and biology can be studied. There is also potential to use neutrons from SNSs to make nuclear waste safe (by "transmutation"), as discussed in Section 12.2.

4.4 Cosmic-Ray Sources

Cosmic rays, which are particles from space which strike the Earth, are discussed in Section 6.1. As regards radiation sources, they are most important as an addition to background radiation. A contamination monitor with a horizontal area of 15×10 cm will measure a background rate of about 150 particles per minute purely due to cosmic rays. This background should not be considered as a pure disadvantage since it can be a useful test of whether the measurement device is functioning, without any need for an artificial radiation source. When measurements are made on radioactive materials, it is important to remember to consider this background rate. Cosmic-ray muons have a high penetration power. they are still detectable deep underground. Because of this, they are ideal candidates for performance tests of particle detectors for charged particles, which are often situated underground.

4.5 Medical Applications

Radiation sources have many applications. One of the main uses of radiation sources is medicine, where they are used in diagnosis and therapy. Indeed, the dominant source of man-made exposure for members of the general public is the application of X, β and γ rays in medicine. Some examples are given for illustration. Taking an X-ray image of the lungs gives a whole-body dose of about 0.03 mSv, and one of the pelvis about 0.5 mSv. Computed tomography (CT) scans tend to give higher doses (5 to 10 mSv for a chest CT), because they involve multiple X-ray images being taken. The procedure with the lowest dose in all of nuclear medicine is an X-ray image of the teeth, which leads to a (whole-body) dose of only 0.01 mSv.

The exposures caused by nuclear-medical tests can be quite substantial, sometimes equivalent to several times the annual natural background dose (about 2.5 mSv). The imaging of the thyroid gland with the iodine isotope ^{131}I leads to an equivalent whole-body dose of 33 mSv. In the past the liver was imaged with the gold isotope ^{198}Au, giving an equivalent whole-body dose of 3.6 mSv. These isotopes have now been mainly replaced, where a γ-emitter is needed, by the technetium isotope ^{99m}Tc (note the "m", marking it as a metastable state (see Section 2.1)). The big advantage of ^{99m}Tc is the appropriateness of its half-life, which is 6 hours. There is enough time (before decay) to prepare a sample of ^{99m}Tc, and for it to be given to the patient and spread around the relevant tissues. However, the half-life is short enough that many of the radioactive decays occur during the examination period and not too many afterwards: this means that the radioactive decays that there are in the patient's body are used efficiently and the total dose required to give good diagnostic results is relatively modest. For example, $^{99\text{m}}$Tc can be injected into the bloodstream, and the location of its decays then show where blood is and is not flowing around the heart. This is called myocardial perfusion scintigraphy. An activity of 1 GBq is normally used, which leads to an exposure of about 8 mSv for the patient. Examinations using PET (see Section 4.2) usually employ fluorine-18 deoxyglucose (^{18}F, half-life 110 min). For oncological (cancer) investigations, 400 MBq of this compound is usually injected, which also gives an exposure of about 8 mSv.

These short-lived isotopes cannot be stored for long periods, and so cannot be transported long distances, because they decay too soon. They must be produced either near the hospital, or even within it. ^{18}F is produced from ^{18}O by proton bombardment according to

$$p + {}^{18}\text{O} \rightarrow {}^{18}\text{F} + n,$$

where the oxygen isotope is distributed in a particular form of water, $H_2{}^{18}\text{O}$ (^{18}O makes up about 0.2 % of oxygen on Earth, and is stable). Since the half-life of the fluorine isotope is relatively long, it can be distributed locally after production at an accelerator. Short-lived positron-emitters, such as ^{11}C ($T_{1/2} = 20$ min), ^{13}N ($T_{1/2} = 10$ min), ^{15}O ($T_{1/2} = 2$ min), ^{82}Rb ($T_{1/2} = 1.25$ min) can only be produced directly in hospitals which have an in-house cyclotron.

"The good news: With your whole-body scintigraphy, no tumour or disease was detected. The bad news: You are now radioactive!"

© by Claus Grupen

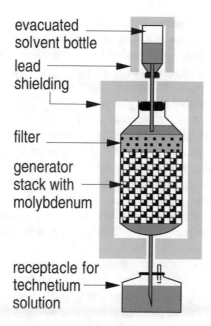

evacuated solvent bottle

lead shielding

filter

generator stack with molybdenum

receptacle for technetium solution

Fig. 4.4 In a moly cow, a solvent (saltwater) passes through a column of insoluble ^{99}Mo, picking up the soluble ^{99m}Tc which is constantly being generated

To produce ^{99m}Tc, the artificial isotope molybdenum-99 is first produced, by striking a uranium target with neutrons. It is then shipped to the hospitals in special units (*moly cows*, see Figure 4.4), which yield the ^{99m}Tc produced in the decay of the ^{99}Mo. This is referred to as *milking*. Similarly, the short-lived γ-emitter ^{137m}Ba can be produced in the laboratory by milking the long-lived cow ^{137}Cs, although this is no longer used clinically.

In many fields of nuclear diagnostics and surgery, the experienced hand of the surgeon is replaced by the tremor-free movements of a robot. This can lead to better clinical outcomes, but also spares the physicians many tiring hours of standing at operating tables wearing lead vests to limit the exposure to X or γ rays.

The highest radiation exposures, by a large margin, occur in the treatment of cancer. Here, there is a compromise to be drawn between the expected therapeutical success and the radiation risk. In tumour therapy, maximum doses for single parts of the human body (i.e. not for the whole body) are extremely high: up to 100 Sv! A whole-body dose over a short period at this level would certainly lead to death from radiation sickness, since the lethal dose for the whole body is about 4 Sv. Cancer patients are able to survive a partial-body dose of 100 Sv only because this dose is applied highly locally and, in general, only one organ is concerned. Furthermore, this high dose is applied in smaller sub-doses of several Sieverts with intervals between them, thereby distributing the exposure over a number of days.

The average per capita exposure in Europe is currently somewhere around 1.9 mSv per year from X-ray diagnosis and about 0.05 mSv per year from other methods in nuclear medicine. This last average value is not completely meaningful because it relates to some very high exposures received only by very few people.

The radiation exposure due to medical examinations varies quite considerably by country. It has to be noted that this exposure has increased over the last few decades. Some experts complain that too many X-ray images are taken. Several thousand tumours in patients are created annually by unnecessary X-ray examinations. A case in point is that taking mammograms (X-ray images to screen for tumours in the breasts) of all women, or doing so very frequently, would be highly questionable. Instead, the application of this useful technique should be restricted to certain higher-risk groups (such as those of certain ages), in which the benefit outweighs the risk. As an example, the UK National Health Service has the sensible policy that women aged over 50 should be screened every three years, and nobody else should unless a tumour is suspected.

There is another area of nuclear medicine which is not directly concerned with radioactive elements. This is hadron therapy, where deep-seated tumours are treated with particle beams, usually either protons or heavy ions like carbon-12. As explained in detail in Section 3.1, these particle beams can be targeted very precisely on a tumour. The number of treatment facilities is increasing substantially worldwide. Figure 4.5 shows a wide-angle view of a treatment room at the Heidelberg Ion Therapy Centre in Germany.

Fig. 4.5 Wide-angle view of a treatment room at the Heidelberg Ion Therapy Centre in Germany. Image credit: GSI/HIT/Siemens

4.6   Radioisotope Batteries

Another important application is radioisotope batteries, which provide energy over a long period of time. They can provide a power density that is six orders of magnitude higher than chemical batteries. They are used if an energy supply is needed over a long time and if the charging of batteries is difficult or impossible, e.g. in medicine or in space missions. Also, radioisotope batteries can be functional under extreme external conditions like very high or very low temperatures. They produce electrical energy from radioactive decays using one of several different methods, such as

- direct charge collection,
- indirect conversion via scintillation processes, or
- thermoelectric energy generation.

Direct charge collection batteries can have very high power density. For example, a battery of this type using polonium-210 can have an energy content of 3 Watt-hours: this is ten thousand times more than that of a normal lithium-ion battery, and the device is of a similar size. In the direct conversion method, the most common type for medical applications, the charge produced by radioactive decay is collected on a spherical metal shell which surrounds the source. In this way voltages of 10 to 100 kV can be obtained. A 100 GBq plutonium-238 source, which is typically used for pacemakers, produces a constant power of several milli-Watts. The patient receives a radiation dose of 1 mSv per year. From the point of view of radiation protection, it is important to dispose of sources (for example after the death of the patient) in a controlled way.

In indirect conversion via scintillation, the ionising particles from the radioactive source strike a material which absorbs the energy of the radiation and converts it into scintillation light in the visible range. This light is then converted into electricity in a photovoltaic cell. Because this is a two-stage process, there is a natural loss of efficiency. However, if the light-emitting properties of the scintillator are well-matched to the light sensitivity of the photovoltaic cell, this loss need not be too large. A major advantage of indirect converters is that in them, the electrics are protected from the radiation by the layer of scintillator, so they can be used with relatively sensitive electrical arrangements.

Many radioisotope batteries are also used for space missions. In this case, the electrical energy is produced via the thermoelectric effect (these batteries are often called RTGs: radioisotope thermoelectric generators). The α particles, which are emitted in ^{238}Pu decay, produce heat in an absorber as they slow down and stop. The heat is then converted into electrical power. A typical ^{238}Pu battery weighs 3 kg and contains an activity of 100 kCi (4×10^{15} Bq). This type of source provides a roughly constant power of 300 W over a lifetime of more than 20 years. The half-life of ^{238}Pu is 88 years, long enough that it should not be the limiting factor on the life of the battery: instead, the limiting factor is mechanical failure, which is very likely to happen before the fuel level has reduced even by 20 %. The launch of rockets carrying radioisotope thermoelectric generators of such high radioactivity presents a certain risk. If the launch of the rocket fails or if the rocket explodes in the atmosphere, there is a clear danger that the atmosphere might be polluted with the ^{238}Pu isotope. There has been at least one example (in 1964) of a military satellite that has burned up, with an RTG breaking up and spreading plutonium, although fortunately this was over the South Atlantic, so there were no significant doses for any people. There have also been at least two cases, including the famous Apollo 13 in 1970, in which an RTG has come down to Earth but not released its radioactive content.

Summary

There is a large variety of radiation sources, with each one being useful for some particular set of applications. Linear and circular accelerators enable us to produce beams of nearly all kinds of charged particles: it is even possible to produce secondary beams of unstable particles. X rays can be generated in X-ray tubes, synchrotron-radiation sources or free electron lasers, depending on the intensity needed. The MeV energy range for photons can be covered by γ rays from radioisotope decays. Neutrons need to be produced indirectly in interactions, for example in radium-beryllium sources. One of the main uses of radiation sources is medicine, where they are used in diagnosis and therapy. The exposures caused by examinations in nuclear medicine can be quite substantial, so it is important to be careful that the benefit of a particular application outweighs the possible risk for the patient, as it often does. Another important application of sources is in radioisotope batteries, which achieve a high power output for their size, and are long-lived.

Chapter 5
What Are X Rays?

*In the end, the public's health is at stake. An underexposed chest
X-ray [picture] cannot reveal pneumonia, and an inaccurate
radiation therapy treatment cannot stop the spread of cancer.*

Charles W. Pickering 1963–

At the end of 1895, Wilhelm Röntgen (there is a drawing of him in Figure 1.2)
was experimenting using a vacuum tube (a glass tube from which the air has been
removed) which had a positively-charged plate at one end and a negatively-charged
plate at the other. It was well known at that time that in this setup, rays, then called
cathode rays, would travel from one plate to the other. It was only later, in 1897,
that Joseph John Thomson discovered that these rays were in fact electrons. Röntgen
observed that, on hitting the positive plate, the electrons produced some penetrating
radiation, which could be detected outside the glass tube by its ability to blacken
a photographic film. He called these new rays "X rays", with the "X" standing for
something unknown.

X rays turned out to be a type of electromagnetic radiation, like light, radio
waves and γ rays, with energies between 100 eV and 100 keV,[1] meaning they have
wavelengths from 10 nm down to 0.01 nm. They are typically produced in X-ray
tubes or transitions of electrons within an atom. A third source, namely synchrotrons,
which produce very bright beams, was detailed in Chapter 4. The X-ray energy range
partially overlaps with that of γ rays: although there is no physical distinction between
a high-energy X-ray photon and a low-energy γ photon, a photon only tends to be
referred to as a γ ray if it comes from a nuclear transition.

X-ray tubes are the most widespread source of X rays, and they start by accelerating electrons using an electric field between two plates: i.e. the electrons start on
a negatively-charged plate (called a cathode), which repels them, and move towards
a positively-charged plate (called an anode), which attracts them. The X rays are
produced when these electrons strike a metal target. Just as in Röntgen's original
design, the target is one of the plates which is causing the acceleration, namely the

---

[1]There are many possible definitions for the precise positions of the edges of the X-ray range. This
definition is provided as an example.

© Springer International Publishing Switzerland 2016
C. Grupen and M. Rodgers, *Radioactivity and Radiation*,
DOI 10.1007/978-3-319-42330-2_5

positively-charged plate (anode). When the electrons strike and enter the anode, two processes lead to the production of photons: bremsstrahlung and vacancy filling, explained below. As a great deal of heat is also produced, the target is usually made of a metal with a high melting point, such as tungsten, and water-cooling is also used (see Figures 5.1 and 5.2).

Bremsstrahlung is the name for the radiation emitted as the electrons interact with the electric fields of the atoms of the metal: these interactions cause the electrons to decelerate and emit photons in the process. The emitted photons have a continuous range of energies. There is no lower limit to the energy of the photons, so even some of the lowest-energy type of radiation, namely radio waves, are given off (although these only make up a tiny fraction of the energy). The upper limit is set by the energy that each electron originally had. The electron-Volt is a very convenient unit for this scenario, because it is defined as the energy that an electron has after having been accelerated in an electric field of one Volt: for example, if the electrons are accelerated with a voltage of 100 kV, they each have an energy of 100 keV when they strike the target, giving a maximum X-ray energy of 100 keV.

Vacancies are created when the incoming electron strikes an atomic electron, and gives it enough energy to leave the atom (ionisation). Within the ion, electrons may

Fig. 5.1 X-ray tube, from the early 1900s. The cathode is on the left, and the anode is on the right

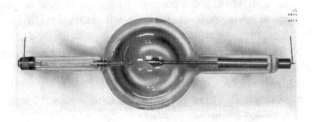

Fig. 5.2 Sketch of an X-ray tube, in the same orientation as Figure 5.1

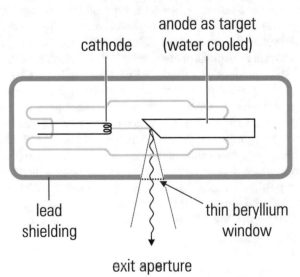

Fig. 5.3 Energy spectrum of an X-ray tube, when operated with a voltage of 65 kV or of 100 kV

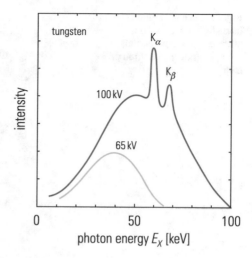

fall from higher states to replace the electron that has been removed. As this higher-state electron falls, energy is released in the form of an X-ray photon. The X rays coming from this process have certain characteristic energies, corresponding to the transitions which are possible within that material. The process is the same as the vacancy filling explained in Section 2.1 (specifically Figure 2.4).

The combination of these two processes means that the spectrum of X rays emitted from an X-ray tube is a continuous bremsstrahlung spectrum with these characteristic lines super-imposed, as shown in Figure 5.3. In that figure, two spectra are shown: one for operation at 65 kV, and one at 100 kV. For the lower voltage, the incident electrons do not have enough energy to ionise the atoms and (indirectly) create the characteristic X rays of the tungsten (which are called K_α and K_β).

The intensity of a beam of X rays is reduced as it passes through matter by the attenuation processes described in Section 3.3. These processes mean that effective shielding can be made using any dense material: lead is a common choice. The intensity of a high-energy X-ray beam is typically reduced to 1 % of its initial value by 1 mm of lead, and lower-energy X rays are attenuated more quickly. Figure 5.4 shows the reduction in intensity of X rays as they travel through a block of lead. The three lines are for X rays produced by accelerating voltages of 50, 70 and 100 kV, and then passed through 0.5 mm of aluminium before striking the lead.

The regulations on the handling of X rays are very similar to the regulations of standard radiation protection. The X-ray regulations in the European Union apply to those X-ray tubes and X-ray installations in which electrons are accelerated to 5 keV or more, as long as there is no acceleration of electrons to above 1 MeV. All installations in which electrons can be accelerated to energies beyond 1 MeV are subject to the regulations of standard radiation protection.

Devices and installations that produce unwanted radiation are exempt from licensing if a dose rate of 1 μSv per hour at a distance of 10 cm from the surface is not exceeded, or if they are given specific design approval (see Appendix B.1).

Fig. 5.4 The reduction in
intensity for X rays from
three different accelerating
voltages as they pass through
lead

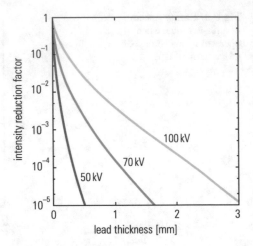

One example of equipment with design approval is old-fashioned TV screens, in which electrons are accelerated up to energies of approximately 20 keV. Modern TV screens, using Liquid Crystal Displays (LCDs) or plasma displays, produce no X rays.

The X-ray regulations, of course, mainly concern X-ray tubes used for X-ray diagnosis and X-ray therapy for humans. It is desirable to obtain the best X-ray image available for a particular radiation exposure. Similarly, it is always desirable to improve the X-ray detection system and image reconstruction, as in this way the radiation dose can be reduced without affecting the image quality. It is highly advisable that the personnel operating the X-ray installations, and also the medical doctors, have the necessary qualifications and experience in the field of radiation protection.

X rays have a broad range of practical applications, and in the following sections, some of the more important examples will be discussed.

5.1   Medical X Rays

The taking of an X-ray image is similar to that of a non-digital photograph, in that the rays hit a film, and cause it to blacken where they are strong enough. The major difference is that the X-rays are observed after they have passed through the person, whereas photographs use reflected light. X-ray images are very useful in medicine because denser tissues (particularly bone) absorb X rays much more strongly than less dense ones, so bones show up as white (no blackening) because they have absorbed all the X rays before they can hit the film. This means that the positions of tissues of various densities in the body can be imaged in a non-invasive way.

In medicine, the patient is rarely given information by the doctor about the X-ray exposure in terms of μSv. If the accelerating voltage, the exposure time, the X-ray-tube current and the geometry of the X-ray imaging system are given, it is in principle possible to calculate an estimate for the dose received.

"Smile!" Cartoon from the Journal 'Life' 1896

Let us take the example of a frontal chest examination. Among the most common X-ray images taken, this tends to be the one with the largest radiation dose delivered. Making some realistic assumptions,[2] a typical whole-body dose can be calculated as approximately $30\,\mu$Sv. The rarer type of chest X-ray image, taken from the side of the patient, gives a larger dose (perhaps $80\,\mu$Sv), because the X rays have further to travel through the body, and so will interact more. It is worth noting that these numbers are comparable to a week's natural background radiation exposure ($50\,\mu$Sv).

Doses from the taking of X-ray images of the hands (see Figure 5.5) or of the teeth are lower, usually below $10\,\mu$Sv. This is in contrast to CT scans (Computed Tomography scans), where the effective dose for a torso scan is between 5 and $10\,$mSv (the dose varies substantially for scans of different parts of the body). The dose, equivalent to a few years' background radiation, is still not high, but it is worth weighing up whether the procedure is worthwhile in marginal cases. The constant improvement in detector technology also allows the doses to fall while keeping the

---

[2]Namely: the X-ray tube operates at $120\,$kV; the product of its current and time is $5.6\,$mA s; the patient is positioned at a distance of $150\,$cm from the focus; the X-ray image covers an area of 35×43 cm^2. The dose-area product is then $0.4\,$Sv cm^2. These numbers give a chest dose of $0.27\,$mSv, implying a $32\,\mu$Sv whole-body dose.

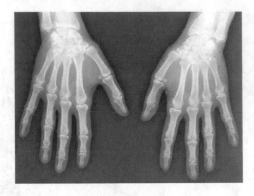

image quality constant. It is important to note that the risk is larger for unborn babies. Therefore great care must be taken when balancing the benefit from diagnostic X-ray investigations for pregnant patients against the potential harm for the unborn foetus.

5.2   X-Ray Crystallography

X-ray crystallography is an important field of application of X rays. When X rays are fired at a crystal, the beam will be spread out and deflected as it interacts with the atoms (the strongest interaction of the X rays being with the electrons in the material). The resulting pattern, specifically the different intensities at different angles in the beam, can be used to infer the position of the atoms, and therefore the internal structure of the crystal. This technique can be applied to many different compounds, since many materials can form crystals: this includes minerals, semiconductors, inorganic compounds, and even organic and biological molecules. The use of X-ray crystallography has enabled scientists to disentangle the structure of many biological molecules, including DNA (Deoxyribonucleic Acid). The large synchrotron facilities (see Sections 4.1 and 4.2), which are sources of very high intensity X rays, are mainly used for uncovering the structure of biological objects, which has many possible practical applications, particularly in the discovery and design of new pharmaceuticals.

5.3   Inspection and Quality Control Systems

X rays provide a useful way to inspect materials, because they can be used to find information about the internal structure of the material, without leaving any significant damage.

A good illustration of the methods used is given by welds. Welds will sometimes contain defects which are internal to the weld, and hidden from view: usually a space (void) in the metal, or some contamination with a different material. These kinds

"I can't see anything!"

© by Claus Grupen

of defects can lead to a deterioration in the mechanical performance of the welded piece, which can be very dangerous, potentially leading to a catastrophic failure.

If X rays are shone onto the welded sections, any irregularities in the joints will show up by their different scattering of the rays. As well as welds, X-ray inspections to find defects are often carried out on tubing, castings, steel plates, plastics, and other components. Modern developments in these inspection devices allow the testing to be performed rapidly, and in a fully automated manner.

Another useful industrial use of X rays is as a thickness indicator. For some applications (including in medicine), it is very important to have smooth steel sheets of very uniform thickness. The thickness indicator works by shining a beam of X rays with a well-known intensity through the sheet, and measuring the intensity of the beam on the far side of the sheet. Small variations in the thickness of the metal cause significant variations in the amount of the X-ray beam which is absorbed, so any deviations in the thickness are made easily visible by this technique.

A further possibility uses X-ray fluorescence. In this technique, a beam of high-energy X rays causes ionisation and excitation in the material, and then de-excitation causes the emission of further X rays, whose energies are characteristic for the material being tested. These machines were first developed in order to scan cargo, with the characteristic energies used to identify explosives and other harmful substances within sealed containers. The use of these systems at borders, ports and airports is becoming increasingly common.

5.4   X Rays in Art

The examination of paintings and sculptures with X rays can reveal interesting details such as the age of a painting, and can also help to identify or verify the artist. It can also be a good (non-destructive) method for the uncovering of forgeries. The Louvre (the French art museum) has installed a proton accelerator to make measurements of this type. In this investigation, the accelerated protons interact with the pigments and the canvas of the painting, ionising some atoms and lifting the electrons of some atoms to an excited state. Upon de-exitation, characteristic X rays are emitted which disclose the identity of the pigments used. This is called the PIXE technique (proton induced X-ray emission). Forgeries can be detected where, for example, a pigment is present which was not in use when the oil painting was supposed to have been created.

Other kinds of artworks (including sculptures) can also be examined in this way, in order to find their chemical compositions. From this, in turn, it is possible to infer information about production methods, and even indirectly about trade routes. This powerful method has uncovered many forgeries and has provided insight into the distribution of various pieces of art all over the world.

There is another technique, very similar to PIXE, in which the artwork is still bombarded with protons, but it is γ rays which are detected: these come from the interactions of the protons with nuclei, and are also characteristic for each material used. This is called PIGE (proton induced γ-ray emission). It can be used in combination with PIXE to reveal more details, as it is more sensitive in some cases.

Hidden pictures beneath oil paintings can also be discovered using the technique of X-ray fluorescence (discussed in the previous section). An X-ray beam causes emission of further X rays of characteristic energies. In this case, the characteristic X rays are used to determine both the identity and position of subsurface pigments. An example of a hidden picture discovered by X-ray fluorescence is given in Figure 5.6. The examination of Vincent van Gogh's painting *Patch of Grass* showed a hidden portrait of a womans's face. Vincent van Gogh could not sell his paintings easily during his lifetime. It is known that he painted over many of his earlier works, just to save money on new canvas.

In the operation of all these systems, of course, it is compulsory to respect the relevant rules for radiation protection.

5.5   Sources of Unwanted X Rays

Equipment which produces unwanted X rays is also subject to the X-ray regulations. This means that if enough unwanted X rays are produced, a licence is required: this is the case, for example, for electron microscopes, high-energy microwave generators and potentially for airport security systems.

Fig. 5.6 A hidden picture of
a womans's face beneath van
Gogh's painting *Patch of
Grass*, discovered by X-ray
fluorescence. (Courtesy:
Kröller-Müller Museum in
the Netherlands (Otterlo).
The image was taken by
chemists from the University
of Antwerp using X rays
from the beamline of the
DORIS III synchrotron
facility at DESY, Hamburg.)

In 2001, there were reports of high exposures for soldiers operating radar equipment in Germany. Radar stations operate by creating and receiving radar radiation: this is a subtype of microwave radiation (with frequencies in the gigahertz range), and as such is non-ionising electromagnetic radiation. It is of course possible for this microwave radiation to create a biological risk for humans if it is at high enough intensity. This potential danger is not discussed here (it is addressed in Chapter 11), however, it has to be mentioned that the generation of radar rays is unavoidably accompanied by the creation of some quantity of X rays. Radar equipment, therefore, produces X rays even though it is not its purpose to do so. In this sense radar equipment is a typical generator of unwanted X rays.

X rays in radar equipment are produced by the use of certain electronic components. In the equipment, devices (of a type called klystrons or another, called magnetrons) are used to accelerate electrons using voltages of 20 to 100 kV. Magnetic fields are then used to make the electrons oscillate back and forth inside a cavity, and in the process produce the radar waves. It is not the intention that these electrons will strike the walls of the cavities, but if they do, their deceleration creates X rays by bremsstrahlung.

In various newspapers, conflicting information was given about the doses received. If the radar equipment is properly shielded, typical dose rates of about 0.06 mSv/h are expected. However, occasionally the radar equipment was not shielded properly or was even unshielded. In such a situation, maximum values of 10 mSv/h have been measured. It should be illegal to enter an area with dose rates this high, except during an emergency: it would count as an "exclusion area" (see Appendix B.1). Other calculations estimate the exposures to have been up to 120 mSv per year from normal operation, i.e. excluding the additional exposures from unshielded equipment during maintenance or calibration work.

An additional risk originates from the use of radium-containing material for displays on the radar equipment. During maintenance and repairs, the radium-containing consoles were cleaned and partially machined. In this way, highly toxic, α-emitting radium dust was very probably released into the air which the operators were breathing. It is estimated that 20000 people operated these radar devices over a period of 25 years. Out of these, 2000 cases of cancer have been reported, of which 200 were fatal, as reported by *Medicine Worldwide*. Using the standard risk factor for cancer incidence of 0.5 % for the given age group and period, one would have expected about 100 fatal cases in a non-exposed population (i.e. in a population exposed only to background radiation and normal background levels of carcinogens). It appears that the exposures during the running of radar equipment significantly increased the cancer-incidence rate.

Summary

X rays are frequently used in medicine and dentistry. They are also widely used in other applications, namely in industrial testing, and in inspection and quality control systems. In solid state physics, the lattice structures of crystals can be determined using X-ray crystallography. Also, in biology, novel X-ray techniques are being used to understand the workings of complicated molecules or even larger molecular structures: X rays were crucial in the discovery of the double-helix structure of the DNA molecule. The X-ray regulations concern X-ray equipment and installations in which electrons are accelerated to energies between 5 keV and 1 MeV. The limits given by the X-ray regulations are defined in a similar way to those in the radiation-protection regulations.

Chapter 6
Is Radioactivity Everywhere?

I am now almost certain that we need more radiation for better health.

John Cameron 1922–2005

Many people believe that radioactivity is all man-made, whether it be made in laboratories, nuclear power plants or nuclear explosions. This is entirely wrong. Since the Earth formed, along with the formation of the Sun and the other planets and moons about 4.6 billion years ago, both it and everything on it have been radioactive to a certain extent. In the distant geological past, the Earth was even more radioactive than it is today. The Sun and the Earth were made from the debris of supernova explosions, in which all the elements of the periodic table, including radioactive isotopes, were created. We are actually made of (slightly radioactive) star dust. Radioactivity is a natural ingredient of all life forms and also of the air that we breathe and the food that we eat. In addition, we are constantly being bombarded by cosmic-ray particles which mainly come either from the Sun or from other sources within our galaxy. These particles constitute a low-level radiation exposure. In the following chapter, the different natural sources of radioactivity will be presented in detail and compared with additional man-made radioactivity.

Natural radioactivity from the environment has three components:

- cosmic rays,
- terrestrial radiation (from the Earth's crust),
- incorporation (eating, drinking, and breathing).

Radiation from these three sources usually affects all parts of the body equally, but a significant exception comes from the inhalation of the radioactive noble gas radon which represents a more localised exposure for the lungs. In addition to these natural sources, there are the exposures due to technical, scientific and medical installations developed by modern society.

© Springer International Publishing Switzerland 2016
C. Grupen and M. Rodgers, *Radioactivity and Radiation*,
DOI 10.1007/978-3-319-42330-2_6

6.1 Cosmic Rays

Cosmic rays are particles coming from space that strike the Earth. Our galaxy is the dominant source of high-energy cosmic rays. The lower-energy particles, which are the majority, predominantly originate from our Sun. Cosmic rays consist largely of protons (about 85 %) and helium nuclei (about 12 %). Only 3 % are nuclei heavier than helium. Nevertheless, all the elements of the periodic table occur as particles in cosmic rays. High-energy electrons are present in cosmic rays, but in much smaller numbers (about 1 %). Photons, in this case γ and X rays, are also incident on the atmosphere in large numbers (for historical reasons, these are not usually thought of as cosmic rays, although they interact in a similar way). Neutrinos, mostly coming from the Sun, enter the atmosphere in extremely large numbers, but effectively play no role in radiation exposures because they very rarely interact with normal matter.

Cosmic rays interact with normal atoms (or, more precisely, with atomic nuclei) of gas in the atmosphere and produce a large variety of particles. The particles produced by the interactions of cosmic rays are called secondary cosmic rays. A very high-energy cosmic ray will produce a cascade, in which the secondary cosmic rays go on to produce further particles (tertiary cosmic rays) in collisions with more gas molecules. The dominant component of cosmic rays which reach the surface of the Earth consists of so-called muons. Muons are particles which have similar properties to electrons with the differences that they are about 200 times heavier and that they are unstable. On average, about one muon per minute strikes each (horizontal) square centimetre of the Earth at sea level. Figure 6.1 shows the changes in proton, electron and muon intensities with depth in the atmosphere. The intensities tend to fall with distance through the atmosphere, as some particles are absorbed as they travel through the air towards the ground. Large numbers of muons and electrons are created in collisions in the upper atmosphere, which is why their intensities rise initially.

Although many more electrons than muons are created as secondary cosmic ray particles, electrons are more likely to be stopped or absorbed in the atmosphere. By contrast, muons are very penetrating particles, which can even be measured at great depths underground.

The exposure due to cosmic rays of a person in Western Europe or in the United States is about 0.3 mSv/yr. It varies with latitude because the Earth's magnetic field shields us from charged cosmic rays to a certain extent, and this points in different directions at different points on the Earth. At the equator, the magnetic field blocks more cosmic rays, and the exposure is lower, about 0.2 mSv/yr. At the poles, the radiation exposure is somewhat higher (about 0.4 mSv/yr), and the large numbers of particles coming down can create fantastic polar lights (aurora borealis and aurora australis). The exposure also varies strongly with altitude in the atmosphere since with increasing height the shielding effect of the air is reduced. For example, the exposure at the top of Mount Everest is 20 mSv/yr while it is about 1.2 mSv/yr in the Alps. At the altitudes of normal passenger flights, the dose rate is about 4 μSv per hour (exposures from flights are discussed in Section 6.5.1).

The radiation exposures of astronauts have been measured with dosimetry telescopes on board the Russian Space Station MIR and the International Space Station (ISS). The exposures are typically 400 μSv per day inside the station, and around

Fig. 6.1 The proton, electron, and muon intensities in the atmosphere, which are smaller closer to the ground. This is the number of particles descending per square centimetre per second and per (solid) angle. The top of the atmosphere is on the left of the graph, and sea level on the right

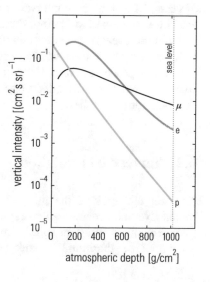

500 µSv for an eight-hour space-walk. About 70 % of this dose originates from the galactic cosmic-ray component and 30 % from protons trapped in the Earth's magnetic field.

By investigating the magnetism present in certain rocks, it is possible to show that the magnetic poles of the Earth have swapped many times in the past. This means that the magnetic north pole becomes a magnetic south pole and vice versa. These reversals occur at irregular intervals, ranging from tens of thousands to many millions of years, with an average interval of approximately 250 000 years. While the poles are in the process of swapping, there is a period (of between a few decades and a few thousand years) when the strength of the Earth's magnetic field is very low, and so it does not shield the Earth from cosmic rays as effectively. In these periods, our planet is exposed to several times the normal level of cosmic-ray radiation. Even so, no particular negative effects on the development of life on our planet have been observed for these times.

6.2   Terrestrial Radiation

The soil and rock of the planet Earth contain substances which are naturally radioactive and provide natural radiation exposures. The most important radioactive elements occuring in soil and rock are the long-lived isotopes potassium-40, uranium-238, thorium-232 and rubidium-87; these radioisotopes will also naturally form part of normal building materials (such as concrete and bricks). Clearly, the level of radiation exposure varies by location depending on the concentration of radioisotopes in the ground. An average exposure from terrestrial radiation for Europe and the United States is about 0.5 mSv/yr. However, this dose exhibits strong regional variations. For example, for the area around the Black Forest in Germany dose rates up to 18 mSv/yr have been recorded. The highest exposures on Earth occur in Kerala on the western coast of India (up to 26 mSv/yr), on the Atlantic coast of Brazil (up to 120 mSv/yr) and in Ramsar in Iran, where doses of 260 mSv/yr have been measured, and there are reports of even higher doses, and yet no increase in cancer rates has been observed. These exposures can be predominantly traced back to high thorium concentrations in the ground.[1]

Areas in which the radiation doses are naturally somewhat elevated provide interesting case studies: large populations have lived over many generations in regions where the exposure is a factor of 10 higher than the average value without any observed disadvantages. The natural radioactivity from the environment was also higher in early geological times (up to 5 times the current level). It is assumed that such relatively high radiation doses were necessary to initiate the development of life on our planet. In the same way, it might be suggested that the radiation from the natural environment was essential to speed up evolution and to create the biodiversity we observe today.

6.3 Incorporation of Radioisotopes

The most important natural isotopes which occur in drinking water and in food are the isotopes of hydrogen (tritium: ^{3}H), carbon (^{14}C), potassium (^{40}K), polonium (^{210}Po), radium (^{226}Ra) and uranium (^{238}U). These natural radioactive elements accumulate in the human body after being taken in by ingestion so that humans themselves become radioactive. The natural radioactivity of the human body is about 9000 Bq, originating mostly from ^{40}K and ^{14}C. This internal radioactivity leads to an average per capita (whole-body) exposure of about 0.4 mSv/yr.

As would be expected, different foods contain different constituents, and so they will necessarily have slightly different levels of natural radioactivity in them. One of the most common foods with a slightly raised level of radioactive content is the banana. Bananas contain high levels of potassium (an essential nutrient), and therefore will contain some radioactive potassium-40. The dose received from eating one normal banana is about 0.1 μSv: there has even been a suggestion that doses should be measured in equivalent bananas eaten! Clearly, in order to get any harmful

---

[1] In the Brazilian state Minas Gerais, there are areas where the mineral apatite contains large quantities of uranium and thorium: so much, that plants grown in these regions, which take up these radioactive isotopes from the soil, will produce an image by themselves if placed on a camera film in the dark!

effects, a person would have to eat thousands of bananas per day, and this diet would kill the person long before the dose had an effect.

In addition to this dose from ingestion, there is an exposure of the lungs due to the inhalation of the radioactive gas radon, which is present in air. ^{222}Rn is produced by the decay of uranium and released through cracks in rocks into the soil, and then from there into the air. If the lung dose created by radon inhalation is converted into an equivalent whole-body dose, the result is an exposure of 1.1 mSv/yr, from the inhalation of this particular radon isotope alone.[2] However, there is at present no evidence that low radon concentrations (under 200 Bq/m^3) induce cancer. The total per capita exposure due to incorporations of natural radioactive substances from the environment amounts to approximately 1.5 mSv/yr.

Like terrestrial radiation exposures, exposures by incorporation vary according to local conditions. Many examples of unexpectedly concentrated exposures come from enclosed areas with poor ventilation and a normal or high level of radon production in surrounding rocks or building materials. The normal concentration of radon in the air outside is about 10 Bq/m^3, but has large variations. A typical value inside well-ventilated houses is 20 Bq/m^3. With poor ventilation, the level in an enclosed space will rise continuously: mines often have a radon level of 100 Bq/m^3, and in some ancient Egyptian tombs and pyramids, which have had little ventilation for thousands of years, radon concentrations of 6000 Bq/m^3 have been found. Figure 6.2 shows a radon monitor used, for example, in mining.

Fig. 6.2 Radon monitor, which measures the radon level per m^3 by detecting α particles. (Image credit: www.industrial-needs.com)

---

[2]Snow samples from Mont Blanc exhibit a radioactivity which is 80 times higher than that of the snow on other mountains in the Alps. This is because there are unusually large numbers of cracks in the (granite) rocks on Mont Blanc, and so the noble gas radon escapes easily into the snow. However, the radon concentration in the snow on Mont Blanc is still harmless.

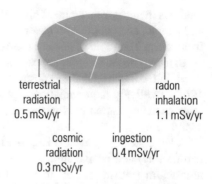

Fig. 6.3 Breakdown of the exposures from the natural environment (2.3 mSv/yr)

terrestrial radiation 0.5 mSv/yr

radon inhalation 1.1 mSv/yr

cosmic radiation 0.3 mSv/yr

ingestion 0.4 mSv/yr

If there are high concentrations of uranium in building materials, yet higher values are possible: record values for radon concentrations in houses were found in the town St. Joachimstal in the Czech Republic. This town was famous for uranium mining with the consequence that in some places radon concentrations of up to a million Bq/m^3 were measured inside houses. The foundations of some of the houses essentially consisted of uranium-containing waste from the uranium mines, which was of course releasing large volumes of radon gas as it decayed. The ICRP recommends that radon concentrations in houses should not exceed 200 Bq/m^3.

Figure 6.3 shows the fractions of radiation exposures by the different components from the environment in a graphical form. A special case is the incorporation of radioisotopes after the Chernobyl accident in 1986 for the relevant parts of the former Soviet Union. There were some significant exposures (possibly up to 10 mSv/yr), because the radioisotopes were more concentrated in mushrooms, venison, and reindeer meat after the accident. The Chernobyl accident is discussed in Chapter 10.

Humans cannot perceive ionising radiation directly, and fortunately low-level radiation is mostly harmless. Nature possibly never saw the need to equip humans with dedicated sense organs for ionising radiation, because such radiation does not present a hazard, and therefore does not require a warning.

6.4   Radiation Exposures from Technical Installations

Radioactive substances are frequently used in scientific and technical installations (for the purposes of this book, hospitals practising nuclear medicine are included under technical installations). However, the dominant source of man-made exposure for members of the general public is the application of X, β and γ rays in medicine, in diagnostics and therapy. There is a wide variation in the doses received for different investigations and treatments, from 0.01 mSv for a dental X ray, to hundreds of millisieverts for certain cancer treatments. In Europe, the average dose per person is about 1.9 mSv per year for X-ray investigations, and 0.05 mSv per year for other methods in nuclear medicine. This will of course vary very strongly between regions and people, and people who never undergo any of these techiques will have no dose from them. The topic of nuclear medicine is covered in more detail in Section 4.5.

In recent decades, other exposures from technical installations have been almost negligible compared to the exposures due to medical diagnosis and treatment. In the 1960s, relatively large radiation exposures were observed from radioactive fallout after nuclear-weapon tests in the atmosphere, but these exposures have now faded away.

A large variety of nuclear isotopes is used in technical installations. For example, tritium (^{3}H) and promethium (^{147}Pm) are used in phosphor screens, the α-emitter americium (^{241}Am) occurs in fire alarms of older types, and plutonium (^{238}Pu), actinium (^{227}Ac), strontium (^{90}Sr), and cobalt (^{60}Co) are used in radioisotope batteries. Radioisotope batteries are discussed in Section 4.6.

Radioactive substances are also used as anti-static materials and filling-level indicators. Phosphate fertilisers contain not only ^{40}K, but also other natural radioactive substances like uranium and thorium. Up to the 1950s, uranium compounds were also used for the production of pigments (with the colours yellow, red, brown, and black), in particular for the glass and ceramic industry. The production of these 'radiating colours' is no longer allowed.

The exposure from smoothly running nuclear power plants is very low (less than 0.01 mSv/yr).[3] In this context it has to be mentioned that coal plants (in addition to the CO_2 emission, which presents a hazard for the Earth's climate) release more radioactive substances into the environment than nuclear power plants in normal operation conditions. The normal level of low-level radiation which is released by coal plants would provoke an outcry in the population if comparable quantities of radioactive substances were emitted by nuclear power plants. For example, in the United States about 800 tons of uranium are annually released into the air by coal plants. However, the use of effective filters can reduce the emission of radioactive substances from coal plants quite substantially.

In Europe, a typical exposure of up to 0.01 mSv/yr per person is produced due to the use of radioactive substances in installations generating ionising radiation (e.g. accelerators) for science and research.

There was one year with an exceptional additional exposure for the public in some parts of the world: this was 1986, after the reactor accident in Chernobyl in the former Soviet Union. In the year after the reactor catastrophe, the average additional exposure in Western Europe may have been approximately 0.5 mSv (and not high enough to be significant in subsequent years). In parts of Eastern Europe, and particularly near the reactor site, doses were of course much higher. This is covered in more detail in Chapter 10.

---

[3]It has, however, to be mentioned that in the vicinity of nuclear reprocessing plants, elevated radiation levels have been measured. For example, close to the reprocessing plant in La Hague, France, ground contaminations of 100 Bq/m^2 of ^{137}Cs and 10 Bq/m^2 of ^{60}Co have been found. In the exhaust air of this plant, ^{85}Kr activity concentrations of several thousand Bq/m^3 have been observed.

Table 6.1 Annual per capita radiation exposure from man-made sources (excluding smoking)

Source of dose	Approximate annual dose (mSv)
Medical X-ray diagnostics	1.9
Other nuclear medicine	0.05
Science and research	under 0.01
Occupational exposure	0.03
Reactor accident in Chernobyl (only 1986, figure for Western Europe)	0.5
Sum (without Chernobyl)	2.0

The total annual average whole-body radiation exposure due to the technical environment is compiled in Table 6.1.

Man-made exposures are dominated by X-ray diagnostics (1.9 mSv/yr). Small contributions come from other areas of nuclear medicine (0.05 mSv/yr), and from science and engineering (0.01 mSv/yr). Occupational radiation exposure (for physicians, from nuclear technology, or at accelerators, etc.) represents only a relatively small fraction (at about 0.03 mSv/yr) compared to the exposures from medical diagnostics. Table 6.2 gives an overview of some typical radiation exposures.

Table 6.2 Typical dose rates or doses for some exposures (whole-body doses)

Type of exposure	Dose or dose rate
X-ray exposure of teeth	10 μSv
Flight Frankfurt to New York	30 μSv
X-ray examination of the chest	70 μSv
Dose limit for general public for discharges from nuclear power plants	300 μSv/yr
Normal smoker	500 μSv/yr
Mammography	500 μSv
γ-ray image of the thyroid gland	800 μSv
Limit for a surveyed area	1 mSv/yr
Heavy smoker (more than 20 cigarettes per day)	1 mSv/yr
Natural radiation	2.3 mSv/yr
Lower limit for a controlled area (see Appendix B)	6 mSv/yr
Positron-emission tomography	8 mSv
Computed tomography of the chest	10 mSv
Limit for radiation-exposed workers in Europe	20 mSv/yr
Limit for radiation-exposed workers in the USA	50 mSv/yr
Limit for emergencies	50 mSv
Maximum worker's dose over the whole life span	400 mSv
Possible future round-trip to Mars (500 days)	1000 mSv
Lethal dose	4000 mSv

Fig. 6.4 Comparison of radiation from the natural environment, exposures from nuclear medicine, legal limits, and exposures from nuclear weapon tests in the atmosphere and from the Chernobyl accident (Western Europe)

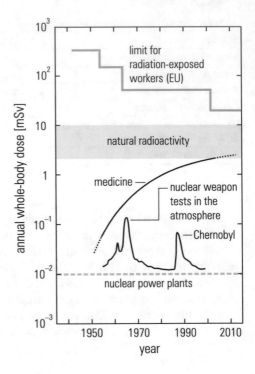

The world average of the whole-body exposure can be estimated to be about 4.3 mSv/yr: about 2.3 mSv/yr from the natural environment and about 2.0 mSv/yr from technical installations (mainly medicine). Depending on location, radiation exposures from natural sources can be quite substantial. In terms of the biological effects, there is no difference between ionising radiation from natural sources and that from technical installations.

Figure 6.4 shows the changes in the contributions to radiation exposure since 1940. The strong increase in radiation exposure due to medical examinations is quite remarkable. The radiation limit of 20 mSv/yr in European countries for radiation-exposed workers is actually quite close to the normal upper limit of the fluctuations of natural environmental radioactivity.

6.5   Specific Environmental Exposures

Some specific locations and practices give additional radiation exposures, which can be surprising. The case of miners (not just uranium miners!) potentially having an additional exposure due to the accumulation of radon has already been discussed. Similar exposures are possible in fields where uranium or thorium derivates are used (for example, as discussed in Appendix A.3, some gas mantles contain thorium).

There are four other notable sources of potentially unexpected exposure, and this chapter ends with discussions of them. They are: air travel, smoking, water, and finally some questionable applications of radioactive sources in medicine.

6.5.1 Exposures from Air Travel

It has already been mentioned that exposures due to cosmic rays rise as we travel upwards through the atmosphere. In particular, inside jets flying at a typical altitude of about 10–12 km, a dose rate of about 5 μSv/h is normally measured. So on a crossing of the northern Atlantic from Frankfurt to New York (6 hours flight time), a flyer receives a total dose of 30 μSv.

There are three factors which will change the exposure that any individual receives from flying. The first is clear: the amount of time flying. The people who fly the most hours are of course flight personnel, so they have the most significant exposures. Indeed, it was airline staff themselves who first pointed out that there is a possible hazard which might even be comparable to exposures in nuclear power plants.

The second important factor is the latitude of the flight, because cosmic ray exposures are much higher nearer the poles, where the Earth's magnetic field shields us from them less effectively. This dependence of the radiation exposure on the latitude is shown in Figure 6.5. A flight from Frankfurt to Tokyo via India gives a dose

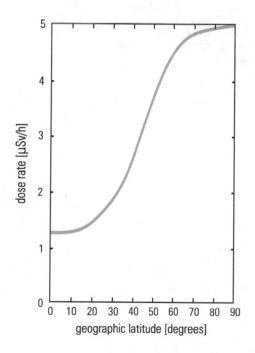

Fig. 6.5 Average dose rate for flights at 10 km altitude, for different latitudes

of approximately 60 μSv. By contrast, a flight from Frankfurt to Tokyo via the polar route (over Alaska) would lead to an exposure of about 100 μSv.

The third important factor is the altitude of the flight, because at lower heights, the cosmic rays have had to travel through more atmosphere, and so more have been absorbed by the air. A Frankfurt-Tokyo flight via the polar route by high-altitude aircraft (for example, Concorde, which used to fly at 18–20 km) resulted in even higher exposures. This was, however, partially compensated by the shorter flight time. Unusual solar activity during the flight occasionally led to an increased radiation exposure. If the radiation level was too high, Concorde descended to below 14 km. The radiation exposure at different heights in the atmosphere, for average geographic latitudes, is shown in Figure 6.6.

Personnel flying at altitudes of 10–12 km are exposed to an annual dose of 2.5 mSv (for 500 flight hours per year). Pilots on high-flying research planes receive a dose of this size in only 125 flight hours. That is to say, flying personnel are working in a radiation-exposed area and the dose rates received should be monitored with suitable equipment.

Flying personnel are also subject to radiation monitoring if their exposures over the course of their flights might exceed a value of 1 mSv/yr. This limit can be reached after as little as 200 flight hours at standard altitudes in the atmosphere. It goes without saying that the exposure should never exceed the limits given by the radiation-protection regulations, e.g. 20 mSv/yr in Europe.

Fig. 6.6 Variation of the dose rate with altitude in the atmosphere for average geographic latitudes (around 50 degrees)

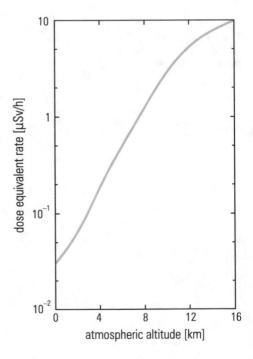

6.5.2 Exposures from Smoking

It is well known that smoking has adverse effects on health, but it is less widely known that radioactive exposures play a significant part in the negative effects. Radioactive isotopes, particularly radioactive lead-210 (^{210}Pb) enter the tobacco plant via its roots from the soil, and also radon (^{222}Rn) will enter the tobacco leaves from the air. This eventually leads to significant exposures of the lungs. The isotopes ^{210}Pb and ^{222}Rn decay after a number of radioactive transmutations into the radioisotope polonium (^{210}Po) ending eventually in stable lead (^{206}Pb).

Smokers are victims of the fact that tar is sticky: these radioistopes tend to stick to the airbourne droplets of tar in the smoke. The contaminated droplets are then inhaled by the smoker, and tend to cling to the inner surfaces of the lung (where there is more tar to help them stick). This means the radioactive substances sit in the lung and decay there, giving the smoker the full dose, whereas a non-smoker would breathe more of the particles out. Then even when not smoking, smokers will get an additional exposure from radon: we all inhale radon as part of normal air, but it is more likely to stick on the tar in smokers' lungs, and decay there, than it is on the lung surface of a non-smoker.

Values in the literature on the radiation exposure of the lungs of smokers vary quite substantially. They span a range of 0.05 Sv up to several Sieverts for the lungs in 25 years of smoking (1 packet of cigarettes per day). If this dose is translated into an equivalent whole-body dose, it means an additional exposure for a smoker

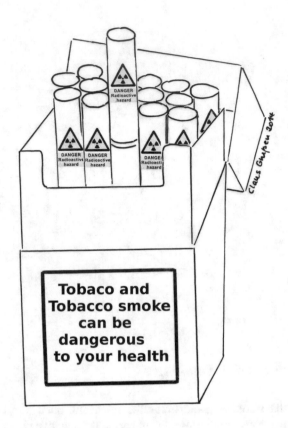

of something like 1 mSv/yr.[4] The polonium content of tobacco varies considerably, so the values given represent only a very rough estimate. There is, however, agreement among physicians that the additional radioactive exposure of the lungs for smokers can lead to lung cancer. This effect has been clearly established for smokers who work in uranium mines, who are of course breathing very radon-rich air (even compared to other miners). For these miners, a substantially increased risk of lung cancer has been demonstrated (compared to non-smoking uranium miners). Clearly, the combined effect of tar, nicotine and exposure to radiation has significant negative consequences. Some scientists comment that smokers manage to find the worst of all worlds, combining the negative effects of chemical substances in the cigarette with the cancer-inducing properties of the ionising radiation from certain isotopes in the tobacco leaves.

---

[4]In the literature there are numbers for the equivalent dose between 0.005 μSv per cigarette and 40 μSv per cigarette. Some authors quote exposures for heavy smokers of 40–400 mSv per year. Such high values can certainly be obtained if the tobacco plants are grown on soil with high concentrations of lead-210 and polonium-210, for example, in certain areas of Brazil and Zimbabwe.

6.5.3   Exposures from Water

Sea water contains 0.01 Becquerels per litre of radon while groundwater holds radon concentrations of 100 Bq/l. At first sight, this is a very surprising difference.

As the name indicates, groundwater originates from the Earth's crust. The rocks making up the Earth's crust contain small concentrations of the radioisotopes uranium, radium and thorium. In the decay of these isotopes, the radioactive noble gas radon is produced. This is washed out by the groundwater leading to an elevated radon concentration. In sea water, the radon concentration is diluted to a large extent. Also, on average, sea water is separated from the nearest rocks (the bottom of the sea) by a larger distance.

Sea water also contains a substantial amount of salt. This salt occurs in the form of sodium chloride and potassium chloride. Therefore, sea water contains the radioactive isotope potassium-40, leading to an activity of about 12 Bq/l. In contrast, this isotope occurs in groundwater with a concentration of only about 0.1 Bq/l.

Rainwater can also wash out uranium, radium and thorium from weathered rocks. These natural isotopes enter into the rainwater and also into the groundwater, and via rivers they also reach the ocean. It is important to note, in particular, that the radon content in rainwater can be quite significant.

The last notable radioactive exposure from water comes from the water itself, in the form of tritium (also called hydrogen-3), which can replace a normal hydrogen in the H_2O molecule. Although it has a half-life of only 12 years, there is always tritium on Earth because it is constantly being generated in small quantities in the atmosphere by cosmic rays. In addition, large quantities of tritium were generated by the atmospheric nuclear-weapon tests in the middle of the last century. Normal drinking water has an activity of about 0.1 Bq/l due to tritium. In order to be a concern, the level would need to be many hundreds of times this number.

6.5.4   Questionable Applications in Medicine

Directly after the discovery of radioactivity, it was suggested that ionising radiation possessed healing powers. As a marketing exercise, this was most widely used by spas in which the water is naturally radioactive.

If certain rock types are present in specific locations, groundwater will move through the rock and pick up significant quantities of radioactive elements. This water will eventually come to the surface in so-called radium mineral springs. There is a variety of possible radioactive elements that can be dissolved in the spring water, including radium, but most of the activity usually comes from radon, which is produced by the radium's decay. The activities of these springs can reach 4000 Bq/l. The idea that these baths were beneficial was so widespread that trips to these spas were even prescribed by doctors. It is known now that swimming in these mineral

Fig. 6.7 Radioactive water
Radithor, containing radium
and mesothorium (now
known to be another form of
radium). This was supposed
to ease all kinds of diseases,
including rheumatism,
headaches, neuralgia, and
constipation

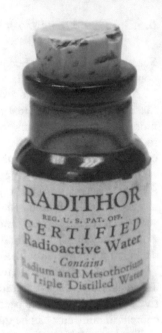

springs or drinking the radium water can lead to considerable radiation exposures.[5]
A famous victim of this kind of treatment was the American golfer Eben Byers, who
died 1932, because he had drunk twelve bottles of radium water (shown in Figure 6.7)
every day, hoping to relieve the pain after an arm injury.

In the mines of Bad Gastein, a spa in Austria, the radon concentration was as
high as $150\,000\,\text{Bq/m}^3$. It has been reported that a research assistant of Otto Hahn,
the discoverer of nuclear fission, was worried about the incorporation of radon and
radioactive dust in his laboratories. When he pointed out this possible hazard to Hahn,
Hahn was supposed to have said: "Don't worry about that; other people pay a lot of
money to go to Bad Gastein and you will get your radon inhalation here for free!"

As well as radioactive spas, some ill-advised radioactive consumer products were
advertised. For example, in France a radioactive hair lotion was put on the market
and the manufacturer recommended its use with the following advertisement:

The most wonderful discovery of the century
radium lotion 'Rezall'.
For the preservation of the hair,
no loss of hair,
no baldheadedness,
no more grey hair!

---

[5]The spas actually advertised these treatments by declaring that the α rays emitted from the radium
effectively massage the cell membranes, thereby improving the patient's health!

This advertisement is quite paradoxical, particularly when one considers that the application of radioactive hair lotion actually leads to the loss of hair.

In Germany in the 1940s, a radioactive "biologically effective toothpaste" with the name Doramad was advertised (see Figure 6.8). The ionising radiation originating from this toothpaste, according to the advertisement, was supposed to massage the gums and refresh the whole mouth.

Even after the risks associated with radioactive substances were well-known, there were some unnecessary uses of them in medicine. For the diagnosis of certain stomach and intestine conditions, it is useful to take an X-ray image after the patient has consumed a so-called contrast agent. The contrast agent absorbs the X rays, and helps provide a clear picture. Non-radioactive forms of iodine and barium are normally used today. Until the 1950s, a thorium-containing contrast agent with the name Thorotrast was used. The thorium from the Thorotrast was predominantly stored in the patients' livers and may have led to cancer of the liver and to liver cirrhosis.

Fig. 6.8 Doramad. The label reads "Radioactive toothpaste" (Image credit: www.mta-r.de/blog/doramad-zahncreme-fuer-strahlend-weisse-zaehne)

Along similar lines, in the 1950s shops selling shoes provided a special X-ray device to test whether the shoes actually fit well, and also radium-containing foot rests with the brand name Elastosan were recommended as major improvement in modern chiropody. These radioactive foot rests exhibited a surface dose-rate of $2.5 \mu Sv/h$, comparable with exposures for flights.

There were even some radium compresses advertised in the period from 1920 to 1960, as an attractive alternative to electric blankets. According to the manufacturer, each radium compress contained a guaranteed amount of at least $100 \mu g$ of ^{226}Ra, corresponding to an activity of 3.7 MBq. This is hundreds of times more than the modern legal limit for this isotope. In addition, because the radium gives off radioactive radon gas, the concentration of radon in a room containing this compress would be hundreds of times more than the ICRP's recommended maximum. With these compresses, the dose rate from γ rays alone was $300 \mu Sv/h$ even at a distance of one metre! The company also promoted and advertised the ease of use of this activator blanket and declared that there was absolutely no danger related to the use of the device. Press reports at the time also stated that if these highly concentrated radium compresses were used correctly, there was no risk of harm.

Summary

Radiation exposures from the natural environment come from cosmic rays, radiation from the ground (terrestrial radiation) and the incorporation of radioisotopes from the natural biosphere. Normally, the three doses are of broadly similar size (with the dose from incorporation being a little higher). The exposure due to technical equipment and installations has its origin almost exclusively in medicine (diagnosis and therapy). As regards its effects, there is no difference between ionising radiation from natural sources and ionising radiation from technical installations.

Chapter 7
What Does Radiation Do to Us?

The technology used to detect if vehicles are carrying radioactive material is so sensitive it can tell if a person recently received radiation as part of a medical procedure.

Timothy Murphy 1952–

Any radiation exposure can potentially have negative effects on health. This can be considered as the basic principle of radiation protection. It is therefore no surprise that damage due to ionising radiation was first observed very soon after the discovery of radioactivity by Becquerel. The biological effect of ionising radiation is a consequence of the energy transfer, by ionisation and excitation, to cells in the body.

The biological effects of radiation absorption are shown in Figure 7.1 in detail. Figure 7.2 shows a very rough classification of different types of radiation damage. Usually they are divided into three different categories (early, delayed and genetic), as discussed in the next three subsections.

7.0.1 Early Effects

This radiation damage occurs immediately after the irradiation, and only appears for high radiation doses. From a whole-body dose of 0.25 Sv upwards, it is possible to see the effects using a blood test (haemogram). For doses of around 1 Sv, clear symptoms of radiation sickness are to be expected. However, the recovery of the patients is nearly guaranteed if sufficient medical care is available. For a whole-body dose of 4 Sv, the chance of survival is 50 %. This dose is called the lethal dose. For a dose of 7 Sv, the death rate is nearly 100 % (see Figure 7.3).

For high radiation doses, the symptoms of radiation sickness occur within a few hours of irradiation. The symptoms are headaches, nausea and vomiting. These symptoms normally disappear after some time. After a quiet period of several days almost without any symptoms, the second phase of the radiation sickness starts. The

© Springer International Publishing Switzerland 2016
C. Grupen and M. Rodgers, *Radioactivity and Radiation*,
DOI 10.1007/978-3-319-42330-2_7

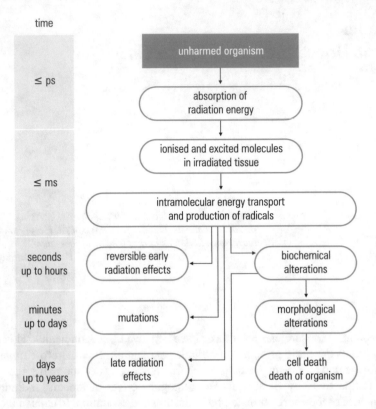

Fig. 7.1 Timings of the biological effects of absorbed radiation energy

symptoms are then fever, haemorrhage, vomiting of blood, bloody faeces and loss of hair. For the highest radiation doses, the quiet phase will be shorter or may even not occur. If the exposed person survives for eight weeks, there is good reason to expect a complete recovery from radiation sickness. However, in some cases death can occur after several months.

Biological tissue has several different repair mechanisms, giving it some ability to rectify damage. Therefore, there is a threshold dose for early effects after irradiation. This means that below a certain dose, no lasting damage is observed. This threshold dose depends on how the dose was distributed in time, and which parts of the body are affected. The smallest value of the threshold is about 0.5 Sv (which is when the exposure all happened at once), and it is closer to 1 Sv if the dose is spread over a longer period. Radiation exposure from natural radiation is certainly far below this threshold. According to current (cautious) thinking there is, however, no threshold dose for the other two effects, discussed below.

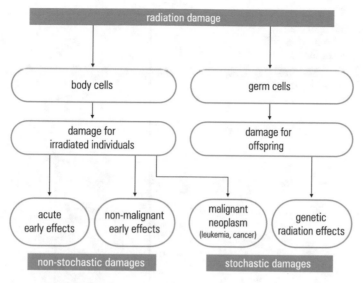

Fig. 7.2 Overview of the different kinds of radiation damage. The early non-malignant effects are symptoms such as temporary reddening of the skin

7.0.2 Delayed Radiation Damage

Delayed (or "late") effects are those which occur after a long dormant period, which can be several decades long. The most common and most frequently-discussed late effect is cancer. In contrast to early effects, whose severity is related to the dose received, delayed radiation-damage effects represent a so-called stochastic risk. This means that the probability of each negative outcome rises with increasing dose, but the severity of each outcome does not (see Section 7.5). The total cancer risk per absorbed dose of 10 mSv is estimated to be to about 5×10^{-4}. This means that out of 10 000 people being irradiated with 10 mSv, on average five of these will later develop cancer due to that exposure.

It is generally assumed that the relationship between the probability to develop cancer and the absorbed dose is a simple straight line (see Figure 7.4). In addition to the assumption of a linear dependence it is argued that there is no threshold for radiation damage: i.e. there is no amount of radiation small enough that it has no damaging effect at all. In combination, these assumptions are called the LNT (Linear No-Threshold) hypothesis, and are quite conservative, as discussed in Section 7.5. Some scientists even assume that humans have no sense organ to warn against ionising radiation because they do not need one since low doses do not present any risk.

It is interesting to compare the cancer incidence as a result of exposure to radiation with other risks. The probability that any person is the victim of a fatal accident (say, a traffic accident, or an accident in the home) in any one year is about 5×10^{-4}, which is comparable to the total cancer risk after a whole-body exposure of 10 mSv.

Fig. 7.3 Early effects: mortality (death rate) after 30 days for different whole-body doses for humans

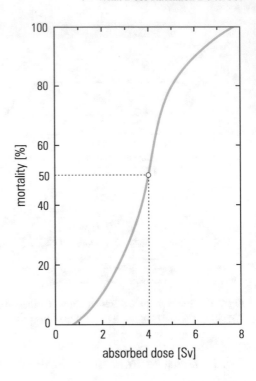

Fig. 7.4 Dependence of the radiation risk on the absorbed whole-body dose in comparison to the 'normal' incidence of cancer (LNT model)

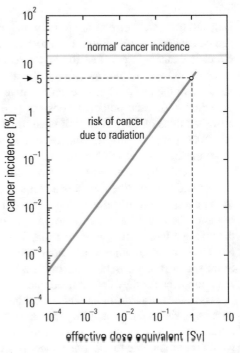

"10 000 nuclei decay per second in our bodies,
and we worry about the price of petrol!"

© by Claus Grupen

7.0.3 Genetic Damage

Radiation absorption in germ cells (the precusors to eggs and sperm) can result in mutations. For the irradiated person, the mutations are not recognisable. They will only manifest themselves in the following generations. During the reproductively significant age of humans (up to the age of 40), about 160 germ-cell mutations occur due to environmental factors. A radiation exposure of 10 mSv will add another 2 mutations on average: this corresponds to less than two per cent of the natural rate of mutations. The average risk factor for these radiation effects is estimated to be 10^{-4} per 10 mSv (i.e. one sufferer would be expected if 10000 people received 10 mSv). This number includes the effect on the first two generations. After this, the probability of transmission is very small.

7.1 Radiosensitivity

The sensitivity to radiation of a piece of tissue is increased if the cells are reproducing rapidly, and also if there is a large variety of different cell types in a small region.

If the cells are dividing rapidly, there is less time available to repair damage before the next division of the cell occurs. If there are many cell types in a small area, it follows that if a few of one type happen to be damaged, there will not be many of that type nearby to replace them. These effects combine to ensure that the time when humans are most sensitive to radiation is as embryos: the cells are dividing rapidly for the embryo to grow, and the nascent organs only contain small numbers of cells. This means that the costs and benefits of perfoming medical procedures involving radiation must be weighed particularly carefully for the case of pregnant women.

It should be mentioned that there are some chemical substances which can modify the biological effect of radiation quite substantially. For example, oxygen, bromouracil and fluorouracil increase the radiosensitivity, i.e. they make human tissue more susceptible to radiation. The water content in the cell also has quite a large influence on the radiosensitivity because potentially damaging free radicals are produced by the ionisation of water molecules. Some carcinogenic (cancer-causing) substances act by increasing the sensitivity of tissue, rather than by causing damage themselves directly.

In the same way that sensitising substances exist, there are also radioprotective substances. For example, mice will survive a radiation dose of 7 Sieverts if they receive an injection of cystamine before the irradiation, while this dose is normally lethal for mice. If an exposure to radiation is fractionated, i.e. received as multiple sub-doses separated by periods of time, its effect is reduced. Clearly, regeneration mechanisms come into play which repair radiation damage between the individual fractions. Also, there is a higher resistance to a large dose if it is preceded by another dose which is small, but significantly above background level ('pre-irradiation'). The use of fractionated irradiation or pre-irradiation has been shown to reduce early radiation damage, and there are some tentative indications that it can reduce the cancer risk.

7.2   Decorporation

It is also possible to remove incorporated radioactive substances from humans by administering suitable drugs. These methods of *decorporation* work by helping the body flush out the incorporated radioactive substance. One possibility is for the drug to bind the atoms of the radioisotope together (chelation), so that the kidneys can remove them in urine. Decorporation drugs can also help by stopping the body absorbing the radioisotope through the gut, so that it passes out in faeces. The best results for decorporation have been obtained with DTPA (diethylenetriamine pentaacetate) and EDTA (ethylenediamine tetraacetate). Figure 7.5 shows the excretion rate of plutonium from the human body after an incorporation caused by an accident. There were two treatments with DTPA (the first very soon after the accident), which increased the excretion rate markedly.

Fig. 7.5 Excretion of plutonium from the human body. There were two treatments with DTPA after an incorporation caused by an accident

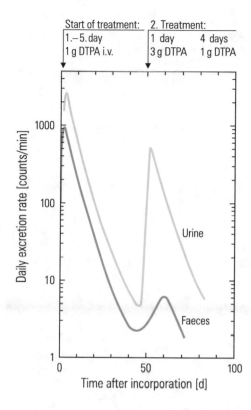

The human body will have some ability to excrete incorporated material, and this ability will vary according to the substance involved. Because this excretion will reduce the amount of the radioactive substance in the body, and therefore its activity, the concept of a biological half-life is useful. This is the time required for the body to excrete half of a quantity of radioactive substance. This is distinct from the (normal) physical half-life. Indeed, both the excretion effect and the decay of the material will act to reduce the activity of the substance in the body over time, so there is a concept of the effective half-life, which is the time for the activity of a particular substance to reduce to half inside the human body. The effective half life is always lower than the other two half-lives. For example, for yttrium-90 (a β-emitter used in radiotherapy), the physical half-life is 64 hours, the biological half-life is 56 hours, and the effective half-life is 30 hours.

Figure 7.6 shows the decrease of ^{137}Cs (physical half-life: 30 years) stored in the bodies of humans and some other mammals. The biological half-life of ^{137}Cs for humans is 110 days, giving an effective half-life of 109 days. The other mammals listed have faster metabolisms, increasing the rate at which the caesium is excreted.

Fig. 7.6 Decrease of the
accumulated ^{137}Cs in the
body for humans and for
some other mammals

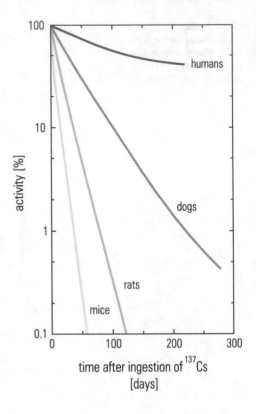

7.3   Non-Human Organisms

Different organisms are resistant against ionising radiation to different extents. For
example, the lethal dose, remembering that doses are inherently per-kilogram measures, for all mammals is about the same (humans: 4 Sv, dogs: 4 Sv, monkeys: 5 Sv,
rabbits: 8 Sv, marmots: 10 Sv). In contrast to that, spiders (with a lethal dose of
1000 Sv) and viruses (2000 Sv) are much more resistant against ionising radiation. If
there were a nuclear holocaust,[1] it would probably only be survived by spiders,
viruses, bacteria and certain types of grass. The idea that cockroaches have an
extremely high radiation tolerance is a myth: although higher than that of humans,
their resistance is similar to that of many invertebrates.

The bacteria deinococcus radiodurans and deinococcus radiophilus can survive
enormous doses (30 000 Sv) because of their extraordinary ability to repair radiation damage. They have even been found in the hot reactor cores of nuclear power
plants. These bacteria somehow manage to repair DNA damage with the help of a
special enzyme system, even if the helix structure of the DNA exhibits about one
million breaks. Deinococcus radiodurans is able to make chemical changes to highly
radioactive waste, which make the process of disposing of it easier and more efficient.

---

[1] Let's hope not!

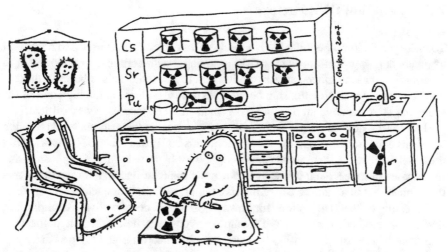

"So, Deino, shall we have some delicious caesium-137 for dessert?"

© by Claus Grupen

For this reason, there is active research into using these bacteria to clean up the radioactively contaminated areas which result from nuclear accidents, military use and earlier generations of nuclear power plants. Because of the high level of resistance to radiation, and also to extreme temperatures, these organisms can survive in meteorites under space conditions over a long time. Consequently they can also propagate over large distances. It is even conceivable that life on Earth was initiated by the impact of meteorites containing such organisms (a hypothesis called 'panspermia').

7.4 Radiation-Absorbing Fungi

The fungus cryptococcus neoformans appears to exhibit the astonishing ability to transform the energy of ionising radiation into usable energy. The fungus was described in detail for the first time in 1976. Its special accomplishment became generally known when it was found after the Chernobyl accident in the sealed nuclear reactor. In general, fungi are rather radiation resistant: they can survive radiation doses of up to 30 000 Sv. However, cryptococcus neoformans goes a dramatic step further: its metabolism increases significantly under irradiation. It appears that the fungus achieves this by using melanin, a class of pigments also common in plants and animals, to derive usable energy from the γ radiation, but the mechanism is not well-understood. In principle, this is a similar technique to that of normal plants: plants transform electromagnetic radiation from the visible range into chemical energy (by photosynthesis), and the fungus seems to be doing the same with radiation from the γ energy range.

As an aside, cryptococcus neoformans can be harmful for humans: it can cause meningitis, especially as a secondary infection for AIDS patients.

7.5   Radiation Risk Factors

To make predictions about the likely effects of an exposure, it is useful to have a set of numbers to represent the cancer risks associated with certain doses. However, in determining these numbers, an assumption had to be made. This is because there is good evidence for the relationship between exposure and risk at the high end of the scale, but the measurement of small risks from small exposures is very difficult.

The solid evidence comes from cancer incidence in the aftermath of radiation accidents and after the dropping of nuclear bombs on Hiroshima and Nagasaki. After these horrific events, there were many people with high, measurable exposures, and many cases of cancer, so it was easy to work out the relationship between them, and to summarise it in a number called the risk factor, different for each cancer type.

To estimate the risk factors for lower doses, the conservative assumption is made that the LNT (Linear No-Threshold) hypothesis holds. This hypothesis is that there is a linear relationship between the absorbed dose and the probability of radiation-induced cancer, and that there is no threshold dose: i.e. no dose is small enough that the risk from it is effectively zero. This means that each additional millisievert of exposure gives an identical increase in risk, regardless of whether the total dose is high or low. The "no-threshold" component is particularly conservative, as the bodies of all living things are known to have some ability to repair minor radiation damage.

Alternative models have been suggested in which the increase in risk with increasing exposure starts slowly, but the risk increases more strongly above a few tens of millisieverts. If this model held, it would mean that an additional millisievert on a low exposure causes little additional damage, but an additional millisievert on a higher dose is more harmful.

Initially, the risk factors were developed based on cancer types with relatively short latency periods, like leukaemia. With the help of computer models of the biochemical behaviour, these results have been extrapolated to other types of cancer. The results of approximations to low doses are, however, somewhat problematic. Additional hypotheses are required to arrive at risk factors which are believed to be reliable for low doses (say, 20–50 mSv), which are of interest for radiation-protection purposes. In the early 1980s, the risk factors determined in this way led to values for radiation-induced cancer of 1.3 % per Sievert. Based on new results, better models and the LNT assumption, the somewhat higher risk factor of 5 % per Sievert is now usually used. This means that if 10 000 people receive a dose of 20 mSv, it would be expected to cause ten additional cancer cases. The risk factors for different types of radiation-induced cancer have also been determined (e.g. 0.5 % per Sievert for leukaemia and 0.85 % per Sievert for lung cancer).

If a radiation worker is exposed to the maximum total dose (in the EU) of 400 mSv over an entire lifetime, there is an additional total risk of 2 % (i.e. of 1000 people working under these conditions, about 20 will develop a radiation-induced cancer). This size of occupational risk appears to be acceptable.

It is possible to use risk factors to illustrate the stochastic nature of cancer risk, namely that an increase in dose increases the probability of each negative outcome, but does not increase the severity of any outcome. Take for comparison cancers of the oesophagus (risk factor 0.3 %/Sv) and of the thyroid (risk factor 0.08 %/Sv). Oesophagal cancer is widely considered to be among the most serious of cancers. In contrast, although it is still a serious medical condition, thyroid cancer is among the least severe cancers, particularly if caught early. A 10 mSv whole-body exposure gives a 3×10^{-5} chance of oesophagal cancer, and a 8×10^{-6} chance of thyroid cancer. A 250 mSv whole-body dose gives 7.5×10^{-4} for the oesophagus, and 2×10^{-4} for the thyroid. The probability of each cancer has risen with the higher exposure (and the probabilities have risen in line with each other), but the potential medical conditions themselves are the same for the two exposures.

Risk factors are statistical, which means that they affect the probabilities of certain outcomes, but some people will be luckier and others less so. There are some good examples of this from the early days of research into radioactivity. Marie Curie, who handled significant quantities of polonium and radium in her laboratory, eventually died of leukaemia. In contrast, Otto Hahn, who dealt with similar quantities of radioistopes, lived until old age. The fact that he did not share the same fate as Marie Curie was considered a miracle. An American colleague visiting Otto Hahn expressed this feeling as: "After having received such high doses from your experiments it is really a shame that you are still alive!"

The additional exposures of smokers to radiation (predominantly because natural radioactive elements from inhaled air stick to the tar within smokers' lungs) are discussed in Section 6.5.2. The equivalent dose for the bronchi of a very heavy smoker (two packets of cigarettes daily) might be as high as 5 Sv over a period of 25 years. This in itself would lead to a risk of a cancer of the lungs or bronchi of something like 5 %. If, in addition, the carcinogenic effect of nicotine and tar is considered, the number is more like 30 %. This high value is obtained because the cancer risks due to ionising radiation and due to chemical effects will reinforce each other.

7.6   Low Radiation Doses

The risk factors for malignant late radiation damage are very low for doses in the range of a few millisieverts, as they must be because background radiation exposures are at least 2 mSv per year. In individual cases it is, for all practical purposes, impossible to establish correlations between an observed sickness and a possible irradiation of this scale, because the normal cancer rate (for people without the additional exposure) is so much higher than the additional risk from the exposure. So the fact that the risk is low makes it hard to observe its scale.

Apart from damage due to ionising radiation, favourable effects after modest radiation exposures have also been observed. This effect is called hormesis. It is suggested that low doses of non-natural radiation might increase the lifetime of cells. The idea is that cells are able to repair minor damage (on the same scale as is caused by natural radioactivity) and that cells become more resistant to damage in general if they are regularly stimulated to repair themselves by being exposed to additional non-natural low-level radiation.

An interesting study was published recently, describing a group of people who accidentally received significantly raised (but still low) radiation doses over a period of multiple years, and appear to have received a health benefit. In Taiwan in the early 1980s, radioactive cobalt-60 (half-life 5 years) was accidentally included in some steel building materials, and these materials were then used to build 1700 apartments. Over the course of two decades, some 10 000 people occupied these apartments, receiving doses between a few millisieverts per year and a few tens of millisieverts per year. After the problem was discovered, the individuals were traced, with the expectation that there would be an increased cancer rate. The researchers found not only no increase in cancer rates, but actually significantly lower incidences of cancer (and also of birth defects). If these results are supported by other findings, then the limits given in radiation protection regulations should be reconsidered, and in particular the LNT hypothesis should be rejected for low to moderate doses.

Nonetheless, for the purposes of radiation protection it must be assumed that any additional irradiation should be avoided if possible.

7.7 Eradication of Insect Pests

A breakthrough in fighting the tsetse fly on the island of Zanzibar was achieved by the so-called sterile-insect technique. The tsetse fly is endemic across large parts of sub-Saharan Africa, and carries a parasite which causes trypanosomiasis (sleeping sickness), a disease which can devastate both livestock and human populations. Using the sterile-insect technique this fly has been practically eradicated on Zanzibar. In this method, tsetse flies are bred in a research laboratory in large numbers, and the male flies are sterilised with low-level γ rays (since it is next to impossible to separate male from female insects, all the flies bred are irradiated). The infertile males are released from planes over the territory. They mate with wild females, which are not then fertilised, and so have no offspring. The release of 8 million sterile male tsetse flies on Zanzibar decreased the rate of trypanosomiasis among cattle from 20 % to below 0.1 %. The success of this sterile-insect technique was aided by the fact that out of a total of 22 different species of this fly, only one type existed on Zanzibar. In addition, the isolated location of an island is particularly well-suited for this kind of endeavour.

The sterile-insect technique has also successfully been used to eradicate the screw-worm fly (cochliomyia hominivorax) in areas of North America. There have also been many successes in controlling species of fruit flies using this technique. This biological method can also be considered when the desire is to control the population of a species rather than eradicating it.

7.8 Metabolism of Plutonium

In the course of the construction of the first nuclear bombs in the United States (the Manhattan Project), the workers building the bombs were exposed to dust particles containing plutonium. Naturally, questions arose about the biological effects of inhaled plutonium. Of the two bombs dropped on Japan at the end of the Second World War, one was made of enriched uranium (mostly ^{235}U) and the other of ^{239}Pu.

There were warnings of potential health risks and it was suggested that a study be undertaken immediately to understand the metabolism of plutonium. A small fraction of the plutonium that had been produced was allocated for animal studies. The plutonium was injected into different animals, and the excretion and retention rates were studied. Since these rates differed substantially for different species, it was difficult to correlate animal excretion and retention data to humans. As a result, there was a proposal to administer small amounts of plutonium to humans to obtain reliable data.

In this context, plutonium was injected into hospital patients at Rochester and Chicago (USA) in the late 1940s. The patients were thought to be either terminally ill, or to have a life expectancy of less than ten years either due to age or to chronic diseases. Different quantities between a few μg and about 100 μg were administered, corresponding to activities of up to 220 kBq. After injection, samples of blood, urine and faeces were analysed at Los Alamos. The physicists and physicians felt reasonably certain that there would be no additional harm to the patients given their preexisting medical conditions.

The urinary excretion data showed a rapid initial excretion rate, although much slower than for radium. This rate levelled off to a constant amount per day after a few weeks. It was found that significant quantities of plutonium were retained in the body in the long term, making the problem of chronic plutonium poisoning a matter of serious concern. Because of this retention, and the significant quantities of plutonium used, the patients received doses of hundreds of millisieverts per year for the rest of their lives.

Out of the 16 patients tracked, ten died within ten years. Four patients survived more than 20 years. Three of the four survivors were examined in 1973, 28 years after the injections had taken place, providing long-term patterns of plutonium retention and excretion. The results of these studies were used as source for estimating permissible limits in the framework of radiation-protection regulations.

Naturally, these experiments with radioactive substances on humans raised serious questions about medical ethics, especially because of the absence of informed consent from the patients selected.

Summary

Biological consequences of ionising radiation are subdivided into early and late effects. Early effects are only observed for doses larger than 250 mSv. In this case, the seriousness of the consequence is directly related to the dose. The lethal dose (50 per cent mortality) is around 4 Sieverts for humans. Late effects (mostly cancer) occur a long while after the exposure, typically 20 years. Here, the severity of the outcome does not depend on the dose, but rather the probability of occurrence of each outcome does. Ionising radiation can also cause mutations in germ cells.

"The radioactive generator not only powers the fridge,
it also sterilises the food!"

© by Claus Grupen

Chapter 8
How Can We Use Radioactivity Productively?

The laboratory technician has succeeded in implementing by means of the atomic pile the Einsteinian principle of inertia of energy.

Gaston Bachelard 1884–1962

Radioactive decay releases energy. As has been discussed in other chapters, the radiation which comes out of the decay deposits that energy in objects and in living beings in a way that can harm them: however, we can also use radiation to provide useful power.

In nuclear power plants, mass is converted into energy according to the famous equation $E = mc^2$. Only in annihilation processes (which cannot drive power plants) does such a conversion have 100 % efficiency. For nuclear transformations like fission and fusion, only a relatively small fraction of the mass can be converted into energy. However, because of the large value of the speed of light, the conversion of a small amount of mass leads to a huge amount of energy. The principle that powers nuclear reactors can best be illustrated by looking at the binding energy per nucleon, shown in Figure 8.1 (the reader will remember that "nucleon" is a term referring to the particles of the nucleus, and that there are two kinds, protons and neutrons).

Nuclear binding energy is the energy required to disassemble an atomic nucleus into separate nucleons. When nucleons bind together to form a nucleus, a small part of their mass is given off as energy. This means that a nucleus is lighter than the sum of the masses of its constituents (if we use the value from when those constituents were separated from each other). So an oxygen-16 nucleus, which is eight protons and eight neutrons bound tightly together, weighs 26.56×10^{-24} g, whereas 8 separated protons and 8 separated neutrons weigh 26.78×10^{-24} g.[1] If we took those 16 separated particles and managed to combine them together, that mass difference (0.22×10^{-24} g) would be given out as energy. That would make 2×10^{-11} J of energy, which is about 120 MeV.

---

[1] 1.6726×10^{-24} g for each proton and 1.6749×10^{-24} g for each neutron.

© Springer International Publishing Switzerland 2016
C. Grupen and M. Rodgers, *Radioactivity and Radiation*,
DOI 10.1007/978-3-319-42330-2_8

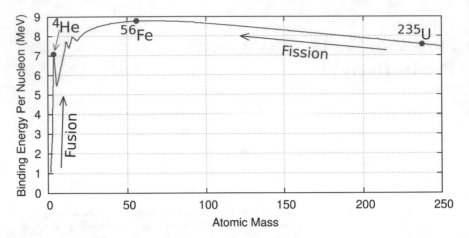

Fig. 8.1 Binding energy per nucleon

Typical binding energies for stable nuclei, such as iron-56 (^{56}Fe), are around 8.5 MeV per nucleon. Heavy nuclei, such as uranium, are less tightly bound, with around 7.5 MeV per nucleon. If a uranium nucleus is split into lighter nuclei, 1 MeV per nucleon is gained, and this energy is given out. This can be looked at in another way: a uranium nucleus is a little heavier (per nucleon) than lighter nuclei. If a uranium nucleus is split up into two lighter nuclei, this mass difference, about 0.1 %, is given out as energy.

In a similar way, mass is converted into energy if four nucleons form a helium nucleus in a fusion process. The mass of the helium nucleus is smaller than the sum of the masses of the four individual nucleons, and this mass difference (about 0.7 %) is converted to energy. Fusion reactions power all stars, and there are attempts to make fusion energy available in nuclear fusion reactors, although all currently-operational nuclear power plants operate using fission. Although the risks can be mitigated, power plants using nuclear fission have the potential to be dangerous, because it is possible for their nuclear chain reactions to run out of control. Also the fission products are highly radioactive and must be stored for long periods in a secure way. On the other hand, fusion reactors never run the risk of a nuclear emergency, and the fusion products, like helium, are not dangerous.

The following chapter introduces the various designs of fission and fusion reactors. This is followed by a discussion of the history of the first ever reactor made by mankind, and then the story of some natural reactors.

8.1 Nuclear Fission Reactors

The most frequently used isotope in nuclear-fission power plants is uranium-235. It breaks up when hit by a slow neutron according to, for example (see also Figure 8.2),

$$n + {}^{235}\text{U} \rightarrow {}^{236}\text{U}^* \rightarrow {}^{144}\text{Xe}^* + {}^{90}\text{Sr}^* + 2n \, .$$

This interaction shows that some neutrons are emitted as part of the fission process (they are "prompt" neutrons). The radioactive fission products created in these reactions will often decay by emitting further neutrons (then called "delayed" neutrons) and/or β^- particles.

The neutrons that come out of this process can go on to cause further fissions, i.e. to repeat the reaction with more uranium-235. In this way, a chain reaction can be set up, where one reaction causes the next, which causes the next. Because two or three free neutrons (on average) come out of a uranium-235 fission, there is the possibility for the reaction rate to increase continuously, with each reaction causing several others. By using materials which absorb neutrons very effectively (control rods), the number of neutrons available to start further fissions can be regulated. This makes it possible to operate a power plant safely at constant power, with (on average) one neutron from each fission going on to create a further fission, so that the process neither increases nor decreases in strength.

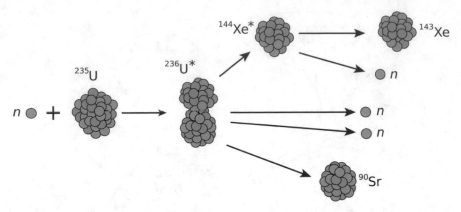

Fig. 8.2 An example fission. The emitted neutrons can go on to cause further fissions of other ^{235}U nuclei. Of the emitted neutrons, the top one is delayed, coming from the ^{144}Xe*, and the bottom two are prompt, coming from the initial fission

In fission reactors, it is ^{235}U which is used to provide the power, because it can be made to undergo this induced fission process sufficiently easily for it to be commercially viable, i.e. the isotope is fissile. However, natural uranium only contains about 0.7 % ^{235}U, with the rest being non-fissile ^{238}U. Surprisingly, it is just about possible to make ^{238}U undergo fission, but the neutrons inducing the fission must have very high energies and be extremely concentrated, so this is useless for power stations.

In order to build a sustainable power plant, it must be sufficiently likely that a neutron leaving a fission will meet another fissile nucleus, which means that enrichment is needed: i.e. the uranium used in the reactor must contain more than the natural proportion of ^{235}U. Enrichment relies on the fact that ^{238}U is slightly more dense than ^{235}U, so when the normal mixture of isotopes is spun in a centrifuge, the isotopes will move so that the part near the axis will contain a slightly higher proportion of ^{235}U. This enriched section is separated off, but because the density difference is very small (about 1 %), many rounds of centrifuging and separation will be needed to make satisfactorily enriched uranium. The part that remains after the enriched uranium has been separated off is *depleted uranium*. Depleted uranium has some military uses due to its high density and relatively low level of radioactivity, but most depleted uranium is simply in safe storage, rather than active use. Different reactors require different enrichment levels, but a typical level is 3 % ^{235}U.

In almost all modern reactors, the fission neutrons which do not go on to cause further fissions are absorbed by control rods, or by the reaction vessel, with the energy they were carrying being turned into heat (useful for the power plant). There is, however, a potential use for the neutrons themselves, rather than just their energy, which is in *breeding*. If the neutrons can be made to attach to ^{238}U nuclei, then ^{239}U

is produced: this rapidly decays into plutonium-239, which is fissile. This is the only practical method for obtaining plutonium. Using this approach, it is possible for a reactor to produce more fissile material than it consumes, even as it produces power. However, because they increase the amount of material available for making weaponry, breeder reactors are politically controversial, and most modern reactors do not breed plutonium. It is also possible to breed other fissile material, such as ^{233}U, which is bred from thorium 232.

The neutrons created in a fission process are of high energy. Since slower neutrons have a higher probability of interaction, the fission neutrons must be slowed down, a process called moderation, so that the chain reaction is maintained.

The moderation of fast neutrons is best done by light materials, because a relatively large amount of energy can be transferred in collisions of neutrons with lighter atomic nuclei. Water (H_2O) is ideally suited for moderation and can be used as a cooling agent at the same time. This provides a very important safety feature because if there is a fluctuation, leading to some additional heat output, some of the cooling water boils, and the fast neutrons from the fission products will no longer be moderated. This reduces the efficiency of the chain reaction, or even interrupts it, so the power output reduces again. Using water for cooling but another material for moderation presents a safety risk: a fission reactor operated with water as cooling agent and graphite as moderator must be considered inherently unsafe. If the water is lost or evaporates, disrupting the cooling, while neutrons continue to be moderated by the graphite (sustaining the chain reaction) this type of reactor can get out of control. This kind of situation can easily lead to a core meltdown. This is exactly what happened in the Chernobyl accident. Reactors using water-cooling and water-moderation have an inherently safer design (this tendency towards very good safety has been demonstrated by the natural reactors discussed in Section 8.4).

There are two main types of reactor using water for both moderation and cooling: boiling-water reactors (Figure 8.3) and pressurised-water reactors (Figure 8.4).

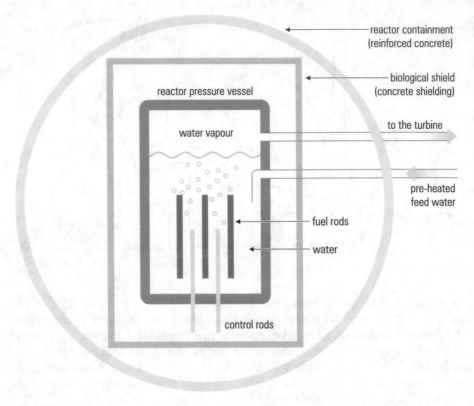

Fig. 8.3 Schematic diagram of a boiling-water fission reactor which is water-cooled and water-moderated

In both cases, the heat generated is used to boil water, and the hot steam then powers a generator via a turbine. In boiling-water reactors, the steam from the primary circuit is used to feed the turbine. When the water boils, it is a less effective moderator, and so the reaction is more efficient towards the bottom of the reactor, where less water has boiled. Because of this, the control rods must be mounted from underneath, so that they are controlling the area where the reaction is most vigorous. This means that if there were an emergency, the control rods would have to be moved up into the reactor core against gravity: if all the power systems fail, this cannot be done, so there is a risk of meltdown. The single system of water also presents a safety issue, because it is not possible to exclude completely the possibility that some contaminated water from the reactor might leak into the mechanical equipment which generates the electricity.

To circumvent this problem, the primary energy can be transferred via a heat exchanger into a secondary water circuit, as in the pressurised-water reactor. The water from the secondary circuit will then feed the turbine without risk of contamination. Pressurised-water reactors have large safety advantages, although the construction of a secondary water circuit is much more complex and also more

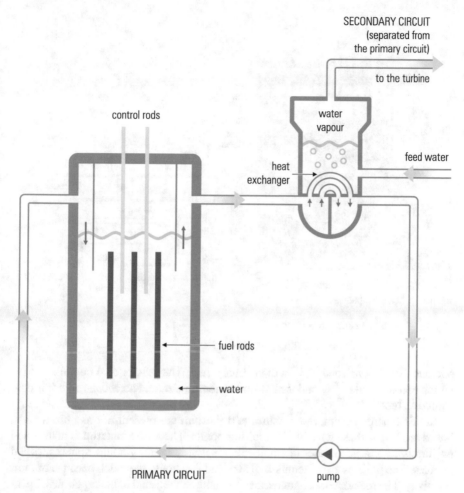

Fig. 8.4 Schematic diagram of a pressurised-water reactor. The water is not allowed to boil in the core, instead it produces steam behind a heat exchanger to feed a generator

expensive. An additional safety characteristic of a pressurised-water reactor is that the water in the primary circuit does not boil (except if there is a fluctuation, when boiling is a crucial safety feature, reducing the moderation and slowing the reaction). This means that the control rods can be mounted on top of the fuel elements. In an emergency, the control rods can be allowed to fall down into the reactor (following gravity) and stop the reaction.

A very different type of fission reactor, called a high-temperature or pebble-bed reactor, may well have a bright future because of its excellent safety features. High-temperature reactors are characterised by their efficient use of uranium and by a high operating temperature (about 1000 °C) compared to boiling-water reactors (about 300 °C). High-temperature reactors use graphite as moderator and helium as

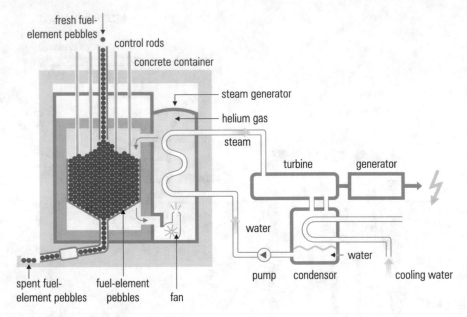

Fig. 8.5 Sketch of a pebble-bed reactor

coolant. Highly enriched ^{235}U is the primary fuel. The reactor also contains ^{232}Th, which breeds fissile ^{233}U under neutron bombardment, and so acts as an additional source of fuel.

In this reactor design, the uranium and thorium are present not as pure metals, but as oxides, as these have higher melting points. The fissile material is distributed uniformly as a large number of small spheres within larger graphite spheres, called pebbles. The pebbles are of tennis-ball size and feel hot to the touch because of their α activity. The reactor core contains several hundred thousand of these pebbles. Spent pebbles can simply be removed from the reactor core at the bottom of the reactor container and replaced by fresh ones at the top without problems, guaranteeing a continuous supply of fuel. Helium gas is an excellent coolant. Neutrons only attach to it extremely rarely, and therefore it does not become activated. The hot helium boils water, and the water vapour drives the turbine. The high operating temperature (about 1000 °C) guarantees an excellent turbine efficiency. It is even conceivable for the hot helium to drive the turbine directly. Figure 8.5 is a sketch showing the design of a pebble-bed reactor.

The major safety advantage of high-temperature reactors comes from the fact that, unlike most fission reactors, a temperature increase in the reactor decreases its reactivity. This provides an automatic stabilisation. There are three (somewhat subtle) reasons for this phenomenon:

1. An increase in temperature causes the graphite pebbles to expand, reducing the uranium density, and with it the reaction rate.

2. ^{238}U nuclei have a certain probability of absorbing neutrons. This probability rises with rising temperature, so then more neutrons are absorbed, and fewer are available to start fissions of ^{235}U.
3. The higher temperatures mean that the neutrons move at higher speed, reducing the probability of each one inducing a fission reaction.

These effects guarantee a negative feedback and an automatic self-stabilisation. These arguments do not hold for 'normal' reactors, since a core meltdown will happen there before the stabilisation temperature is reached (uranium melts at 1130 °C).

The self-stabilisation in pebble-bed reactors works because the ceramic graphite pebbles can withstand much higher temperatures (2000 °C). Because of the negative feedback in pebble-bed reactors, the maximum operating temperature stabilises at 1600 °C, removing the possibility of a core meltdown, at least in theory. This point means that, unlike for Chernobyl-type reactors, it is not a concern that the coolant and moderator are not the same material. A pebble-bed reactor could in principle even function without neutron-absorbing control rods. Instead, the operation temperature could be controlled by the flow rate of the coolant. Nevertheless, control and safety rods will be employed for an extra layer of safety. They will also be used for reactor shutdown.

High-temperature reactors produce much less power per cubic metre (around $6\,\text{MW/m}^3$) than boiling-water or pressurised-water reactors (typically $60\,\text{MW/m}^3$). This gives high-temperature reactors the slight disadvantage that the reaction vessel must be larger, but there is the significant advantage that, with the heat more spread out, it should be impossible for the fuel to rise above its melting temperature. Nonetheless, as an additional safety measure, sophisticated cooling systems will be installed.

One problem of fission reactors in general is the processing and storage of nuclear waste (discussed in more detail in Chapter 12). In nuclear fission, enormous amounts of solid and liquid radioactive waste are generated. The liquid waste is usually stored in large tanks. The storage of the radioactive material would be much easier if it could be concentrated.

As an example, strontium-90 is a particularly harmful component of nuclear waste with a half-life of about 30 years. If incorporated, it can replace calcium in bones and so be retained in the body for long periods (it is often called the 'bone seeker'), causing a significant dose. Work is currently being done on a new set of materials called Octahedral Molecular Sieves, which can extract radioactive strontium ions from solutions with efficiencies up to 99.8 %. New technologies such as these could help to clean up nuclear waste.

8.2 Fusion Reactors

Fusion reactions provide the energy that makes the stars shine. If this source of energy could be made available on Earth, all energy problems would be solved. In the 1930s, hydrogen fusion was discovered to be the energy source of the stars. There

is a popular anecdote about the discovery of the principles of the hydrogen fusion chain: Carl Friedrich von Weizsäcker was taking a walk with a girl on a nice summer night. The girl, who later became his wife, remarked on how beautifully the stars were shining. "Yes," said Weizsäcker, "and right now I am the only one who knows why." Sometimes this anecdote is also attributed to Hans Bethe.

The Sun is a ball of plasma: gas in which the electrons have been stripped away from the atomic nuclei, and those components move freely and independently. It is in this state that nuclei can fuse. The main fusion mechanism in stars of the size of the Sun is based on a sequence of three reactions. Firstly, two protons collide and interact, making a ^2H nucleus and a positron (and a neutrino). The ^2H, i.e. one proton and one neutron together, is called deuterium, and is written here as d (similarly, later we will write ^3H, i.e. one proton and two neutrons together, as t, and call it tritium):

$$p + p \rightarrow d + e^+ + \nu_e .$$

Secondly, that deuterium nucleus is hit by another proton, and the result is ^{3}He and a γ ray (^{3}He is two protons and one neutron together):

$$p + d \rightarrow {}^3\text{He} + \gamma.$$

Thirdly, after the above pair of reactions has happened twice, the two ^{3}He nuclei come together, making the final product, ^{4}He, and releasing two protons:

$${}^3\text{He} + {}^3\text{He} \rightarrow {}^4\text{He} + 2p.$$

In addition to this main mechanism, there are other, rarer processes yielding smaller quantities of lithium, beryllium and boron.

The two particles ($p + p$, $p + d$, etc.) which might collide and fuse are both positively charged. This means that they repel one another, and so they need to have very high energy (or equivalently very high temperature) to get close enough to fuse. The chances that they are brought close enough to fuse are also raised if there are large numbers in a small volume for an extended period. In short, fusion requires three important conditions:

- high temperatures,
- high plasma densities, and
- sufficiently long confinement times of the plasma.

Clearly, these three conditions are met within stars.

Fusion reactors on Earth use a slightly different process, deuterium-tritium fusion:

$$d + t \rightarrow {}^4\text{He} + n.$$

Because the deuterium and tritium nuclei are larger than protons, but have the same charge, it requires less energy for them to get close enough to fuse. This means that it is easier (although still very difficult!) to create the conditions for this fusion to take place on Earth than it would be for the hydrogen fusion described above.

In the deuterium-tritium fusion process, the unstable compound nucleus ^{5}He is initially formed, which quickly decays into ^{4}He and a neutron. The α particle (helium nucleus) leaves with an energy of 3.5 MeV and the neutron gains 14.1 MeV. The α particle has a very short range and will deposit its energy in the immediate vicinity, heating up the plasma. If the heating by α particles is sufficient to compensate for the leakage of energy out of the plasma, a self-sustaining fusion reaction can be maintained. The neutron will escape from the plasma.

Deuterium can be extracted from sea water by electrolysis. Tritium can be bred by firing neutrons at lithium, using the reaction:

$$n + {}_3^6\text{Li} \rightarrow {}_2^4\text{He} + t.$$

If the reaction plasma is surrounded by lithium blankets, the neutrons emerging from the fusion can breed further tritium nuclei.

There are two fundamentally different proposals to maintain the fusion process on Earth in a controlled fashion. The technique of inertial fusion is predominantly being followed in the United States while fusion by magnetic confinement is mainly being tested and investigated in Europe.

8.2.1   Inertial Fusion

The technique of inertial fusion, also called laser fusion, uses hollow spheres made of plastic, and typically 1 mm in diameter, to carry the fuel. These pellets are filled with a mixture of deuterium and tritium at high pressure and are then cooled down to extremely low temperatures. This causes most of the deuterium-tritium mixture to freeze as thin solid coating on the inner wall of the sphere, with a small fraction remaining as a gas within the sphere.

To run the fusion reactor, these deuterium-tritium pellets are injected into a target chamber, where they are bombarded with intense laser beams. The large energy deposition from the laser pulse evaporates the plastic shell of the pellet, which expands rapidly outwards. This causes a pressure in an inward direction which accelerates the deuterium-tritium layer towards the centre of the pellet. That part of the deuterium-tritium gas which had remained at the centre of the pellet will be compressed by the formerly solid deuterium-tritium layer, which is streaming under high pressure towards the centre. In this process, temperatures over 100 million °C are produced for a short period, and this is sufficient to initiate the fusion process. This fusion will then provide enough heat and pressure, for a very brief period, to cause the formerly solid deuterium-tritium layer to fuse as well. In order to obtain a commercially viable amount of energy, these reactors will have to process a continuous stream of the deuterium-tritium pellets within a reaction chamber. Figure 8.6 is a sketch of a power plant based on inertial fusion.

It is possible to generate laser beams of extremely high power density. The lasers at the Lawrence Livermore National Laboratory Ignition Facility reach 500 Tera-Watts for about a nanosecond. For comparison, the USA uses (constantly) about 3 Tera-Watts of power. Each laser pulse provides a maximum energy of 1.9 Mega-Joules. This is about as much energy as is needed to bring 6 litres of water from room temperature to 100 °C.

A fusion reactor based on inertial fusion is a strong source of neutrons. Neutrons can travel many centimetres through matter, and it is unavoidable that some radioactive isotopes are formed when neutrons are absorbed in the surrounding material (this is called activation: see Section 3.2). By careful selection of materials which have low probabilities of neutron activation, and short decay times when they have been activated, the production of radioactive waste can be limited: it is expected that fusion technology will produce far less radioactive waste than fission.

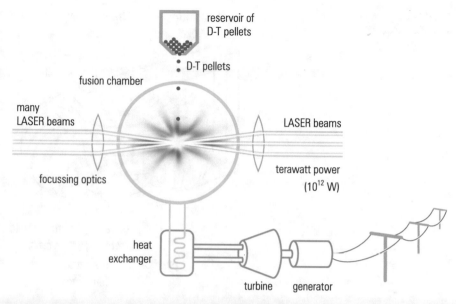

Fig. 8.6 Sketch of a power plant based on inertial fusion

Any hot matter emits radiation, with the radiation being higher-energy for hotter matter. For a glowing bar of iron (at 800 °C), for example, the radiation is in the form of visible light and infra-red. The hot plasma, at a temperature of around 100 million °C, emits radiation in the form of X rays. The X rays take away about 10 % of the fusion energy, and go on to interact with the material of the reaction chamber. In these X-ray interactions, many small fragments of the coating of the chamber walls can be detached, polluting the target chamber. The frequent neutron impacts will also cause some brittleness of the reactor materials. Work is ongoing to overcome these technical hurdles.

8.2.2 Fusion by Magnetic Confinement

In this type of fusion reactor, a high-temperature plasma is produced, and then stored by magnetic confinement over a longer period of time. For this technique, temperatures of about 100 million °C, confinement times of about a second and plasma densities of over 10^{20} particles per cubic metre are required.

Energy losses from the plasma are mainly caused by radiation. The radiation loss of a plasma is related to its surface area, while the energy content of the plasma depends on the volume. This means that the energy retention improves as the size of the plasma increases, because a larger region of plasma has a larger volume, in relation to its surface area, than a smaller one. The plasma confinement is obtained

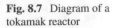

Fig. 8.7 Diagram of a
tokamak reactor

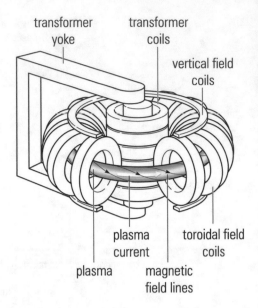

by a rather sophisticated arrangement of magnetic fields: the goal is to store the
plasma in a closed torus (doughnut shape). To confine the charged particles into this
shape, a magnetic field is produced which causes the plasma particles to move on
spiral orbits inside the torus. Furthermore, an additional magnetic field causes the
plasma to pinch, i.e. to be compressed and contained by magnetic forces, keeping it
away from the walls of the chamber (see Figure 8.7). The chamber, called a tokamak
(which is an abbreviation of a Russian phrase), must be completely sealed to the
air, as even small numbers of air molecules would pollute the plasma and stop the
reaction.

The heating of the plasma can be performed in various ways, and in most cases
all four of the following methods will be applied:

- A powerful transformer with a high current will induce a high current in the plasma.
  The current will leave heat behind as it dissipates. This method is called ohmic
  plasma heating.
- Deuterium and tritium ions are first accelerated in a linear accelerator, up to energies of typically 300 keV each. The injection of these energetic particles heats
  the plasma. The magnetic field prevents any charged particles from entering the
  plasma, so the ions are neutralised by the addition of electrons before being
  injected.
- In the same way, electromagnetic radiation can heat up the plasma, as long as the
  frequency of the radiation is chosen carefully so that it will be absorbed efficiently
  by the charged particles of the plasma.
- Finally, some of the reaction products of fusion, specifically the α particles produced, will be retained in the plasma leading to further heating.

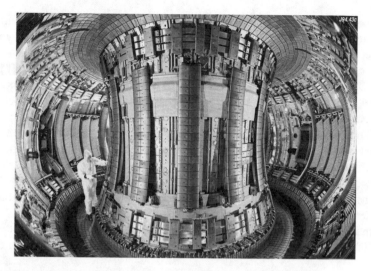

Fig. 8.8 Photo of the interior of the Joint European Torus (JET). The dimensions of the fusion reactor can be estimated from the size of the technician in the left-hand part of the photo. (Photo credit: JET Culham, England)

In very much the same way as in inertial fusion, the neutrons produced in the fusion process will escape from the plasma. These neutrons have substantial energy: they present the basis for the energy production of a fusion reactor. Figure 8.8 shows a photo of the plasma chamber of a reactor built on the principle of tokamak fusion. The follow-up project after the JET fusion facility is the International Thermonuclear Experimental Reactor (ITER) at Cadarache in France. ITER is a joint European-American-Japanese-Russian fusion project to demonstrate the feasibility of producing fusion energy economically. The American participation was, however, significantly reduced in 2008 due to financial cuts to this field of science.

10 g of deuterium, which can be extracted from 500 litres of water, and 15 g of tritium, which can be produced from 30 g of lithium, contain about 3×10^{24} atomic nuclei each. Per fusion process, one neutron with energy 14.1 MeV is produced. If the energy of the neutrons can be transformed into the production of electricity, about 2 million kWh will be obtained. This energy is sufficient to cover the energy consumption of a single person over his entire life.

Apart from the neutron-activated reactor material, fusion reactors produce no nuclear waste. In addition to the avoidance of nuclear waste, the excellent safety of fusion reactors is a very important point. In contrast to fission reactors, a fusion reactor can never suffer a severe radiation accident as happened in the Chernobyl disaster. The conditions for the fusion of the nuclei must be maintained constantly with strong magnets in a precise balance. Any disturbance causes this balance to disappear, and the reaction to stop immediately. This means that it is impossible for an internal fluctuation or an external event (such as a plane crashing into the reactor) to cause sudden uncontrolled fusion.

On the face of it, it might appear that fusion power plants should have difficulty containing their energy, because the most powerful nuclear weapons employ fusion. However (as explained in Chapter 9), a fusion bomb can only be ignited using the extreme conditions of a fission explosion, and even then only with exceedingly precise control. These conditions are not possible in a tokamak. It is also worth noting that a nuclear bomb thrown in a terrorist attack onto the fusion reactor cannot cause the fusion reactor to get out of control: the breach of the chamber wall would simply allow an inrushing of air, the (non-explosive) escape of the fusion fuel, and the destruction of the balanced magnetic fields. This means that the reaction would stop immediately. This extreme sensitivity of a fusion reactor is an important safety feature, because it will prevent the reactor running out of control in any emergency situations. In such situations, the reactor will simply stop working without polluting the environment.

These safety aspects are a strong argument for fusion reactors. It appears that it will someday be possible to master the solar furnace and use it on Earth, and therefore to avoid the possible risks which are related to the production of nuclear energy by fission.

8.3   The First Nuclear Reactor

On 2nd December 1942, the first nuclear reactor came into operation. Enrico Fermi and his colleagues succeeded in maintaining a self-sustaining chain reaction in a graphite-moderated reactor with uranium oxide as fuel. The reactor (shown in Figure 8.9) was set up under the grandstand of the squash court on the site of the sport stadium of the University of Chicago. This reactor was constructed from blocks

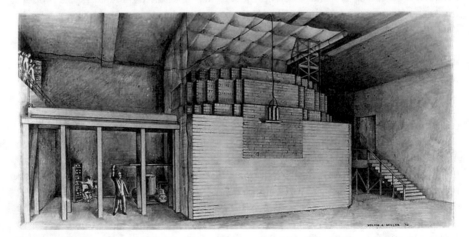

Fig. 8.9 Sketch of the historic reactor in Chicago, in which the first controlled, self-sustaining chain reaction was maintained. (Image credit: Argonne National Laboratory)

of uranium oxide and blocks of high-purity graphite, arranged in a three-dimensional grid pattern. The reactor was meant as an experiment to demonstrate the feasibility of a self-sustaining chain reaction. There was no cooling.

Before the reactor was built in Chicago and before it was set into action, thirty small-scale reactors had been tested to measure the neutron yield and to determine how much uranium was needed for a reactor to have a self-sustaining chain reaction (the critical mass). The reactor was equipped with a three-fold safety system. Firstly, there was an automatic system consisting of cadmium control rods. Secondly, there was a hand-operated emergency safety rod, which was connected to a rope held by a physicist. The rope could simply be dropped or cut with an axe if it should turn out to be necessary, e.g. if the automatic shutdown system was not working properly. The third security system consisted of a team of technicians, who were standing by, and would flood the reactor in an emergency situation with a neutron-absorbing cadmium-salt solution, should the automatic shutdown system and also the hand-operated safety-rod system fail.

The chain reaction was initiated by withdrawing the cadmium control rods slowly out of the reactor. While this was being done, the neutron production was measured for different positions of the control rods. When the neutron rate increased exponentially, it became obvious that a self-sustaining chain reaction had developed in the reactor. In this way it was demonstrated that a controlled, self-sustaining chain reaction in a uranium reactor would definitely work.

8.4   Natural Nuclear Reactors

1.7 billion years ago, a set of natural reactors started to operate in Oklo in Gabon (Central Africa). At over a dozen sites, reactors produced energy, with some interruptions, over a period of several million years (see Figure 8.10). In 1970, French scientists found a very unusual ratio of isotopes of uranium in a mine in Oklo. On Earth, in meteorites and in lunar rock (and indeed throughout the whole solar system) the uranium isotope ^{238}U occurs with an abundance of 99.3 %, and ^{235}U with an abundance of 0.7 %. Other uranium isotopes (such as ^{234}U) are very rare in comparison. This present isotopic ratio, 99.3 to 0.7, is consistent with the assumption that these two isotopes were of equal abundance during the formation of our solar system. The isotopes have different half-lives (^{238}U: 4.5 billion years, ^{235}U: 0.7 billion years), and so the ^{235}U has decayed more quickly over time, leading to the ratio currently observed.

It came as a surprise that the isotopic abundance of ^{235}U was only 0.35 % in the Oklo mine. Comparable levels of ^{235}U are also found in spent fuel rods from modern nuclear power plants. Since the observed isotopic abundance of other elements also corresponded to those which are found in nuclear power plants, it was natural to deduce that a natural nuclear fission chain reaction had been running in the Oklo mine. By extrapolating back, using the current global ratio of abundances, it is possible to calculate what the abundance of ^{235}U would have been before any reactions at Oklo.

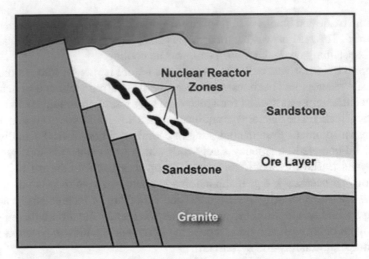

Fig. 8.10 Sketch of the uranium-containing deposits held in position in the surrounding geological formation over a long period in the Oklo reactor. (Image credit: US Department of Energy, Office of Civilian Radioactive Waste Management, Yucca Mountain Project)

The figure is 3 % ^{235}U, which is similar to the ratio in reactor-grade low-enriched uranium.

The rocks in the Oklo mine are very porous and they have many cracks. They are therefore very permeable to rainwater. Rainwater entering the uranium-containing mineral is an ideal moderator for neutrons. Due to spontaneous fission, or also from cosmic-ray interactions, there are always a few neutrons available which can initiate nuclear-fission processes. Neutrons created in fission processes are relatively fast, but at Oklo they were moderated by the rainwater and initiated a chain reaction, and sustained it for a while. The energy created in this way led to the evaporation of the water, thereby interrupting the chain reaction, because the fast fission neutrons were no longer being moderated. Only when further water entered the rock, or when the evaporated water condensed, did the chain reaction restart. As well as by this boiling of the water, the natural reactors were certainly also modulated by the alternating rainy and dry seasons in Central Africa at the time.

It has been estimated that the Oklo reactors reached a thermal output of about 100 kW. From the fission products, it is possible to calculate that several tons of uranium were processed and also a substantial quantity of plutonium was bred. Of course, the Oklo reactors stopped working when the abundance of ^{235}U fell below a certain critical limit. The fission products of the natural reactors and the heavy elements bred (e.g. ^{244}Pu, half-life 83 million years) migrated only a few metres from where they had been created. They stayed essentially where they were produced. They were also not distributed or washed away by the groundwater. In this experiment, which cannot be repeated in a laboratory, nature demonstrated that over a period of 1.7 billion years, certain geological formations are suitable as a natural repository for radioactive waste.

At the time of writing, no prehistorical natural reactors other than the ones at Oklo are known. Because of the decay of ^{235}U, it is not expected that there could be any natural reactors operating today.

Summary

Nuclear fission has led to the development of (civilian) nuclear power. The transformation of mass into energy can be realised by fission of heavy elements but also by the fusion of hydrogen to helium, the energy production mechanism which makes the stars shine. There are a number of different ways to put nuclear fission and fusion into practice. Carefully designed fission power plants based on water-cooled, water-moderated systems or on pebble-bed reactors guarantee a high degree of safety. The design, implementation and specific construction of a fusion reactor is still a matter for research and development. In the long term, mankind's energy demands are likely to be satisfied with the help of nuclear fusion.

Chapter 9
How Can We Use Radioactivity Destructively?

There is no direct evidence that nuclear weapons prevented a world war. Conversely, it is known that they nearly caused one.
Joseph Rotblat 1908–2005

The discoveries that led to the development of nuclear weapons took place in the 1930s. In 1938, Otto Hahn and his assistant Fritz Straßmann were trying to produce transuranic elements (elements heavier than uranium, which is the heaviest in nature) by firing neutrons at uranium. To their surprise, they found that most of the products were lighter elements, like caesium and barium. This showed that, far from building up the nuclei to heavier elements, the neutrons had caused the uranium nuclei to break apart, in a process now called fission. Within a few months, Lise Meitner and her nephew Otto Frisch, a chemist, had explained correctly what had happened. Lise Meitner was not impressed by the physics expertise of Otto Hahn (whose surname means "chicken"), and she was reported to have said, "My little chicken, you understand nothing about physics!"

It soon became clear that the component useful for fission was the uranium-235, which makes up about 0.7% of natural uranium (the rest being uranium-238). Calculations showed that as little as 5 kg of pure ^{235}U could make a powerful bomb with a destructive power equivalent to several thousand tons of dynamite. However, obtaining pure ^{235}U by enriching natural uranium required sophisticated techniques, as explained in Section 8.1.

In the context of the Second World War, which was starting as these discoveries were being made, it was natural to try to use this destructive force for military purposes. Within fifteen years of the discovery of fission, two fundamental designs of nuclear weapon had been developed: the atomic bomb, and the hydrogen bomb. These two designs mirror the two types of power plant that were discussed in Chapter 8.

An atomic bomb, also known as a nuclear bomb or a fission bomb, uses two or more shaped pieces of highly enriched uranium (usually around 95% ^{235}U) or of plutonium-239. Each of these pieces is, on its own, not sufficiently large to sustain a fission chain reaction: they are "subcritical amounts". The pieces are suddenly

© Springer International Publishing Switzerland 2016
C. Grupen and M. Rodgers, *Radioactivity and Radiation*,
DOI 10.1007/978-3-319-42330-2_9

brought together by a mechanical implosion, and a chain reaction starts immediately (the material goes supercritical), and this causes a devastating explosion. The idea of critical mass (the amount of fissile material which can sustain a fission reaction on its own) is more important in weaponry than in power production because in the weapon, criticality cannot be maintained using moderators, so the neutron production must be much more concentrated. This is the reason why such very high enrichment levels are needed.

In the process of research into nuclear weapon design, one proposal was to use an antimatter explosion, rather than a purely mechanical one, as a trigger. In this case, they were considering using anti-hydrogen, which is composed of a positron circling an anti-proton. There were two problems with this. Firstly, because antimatter will annihilate and produce energy whenever it makes contact with normal matter, it is very hard to store. Secondly, the only way to produce antimatter is to make each of its components (positrons and anti-protons) in a particle accelerator, and then slow them down and get them to interact to make neutral anti-atoms. This is a very challenging process, and at the time of writing, only about a million-billionth of a gram of anti-hydrogen has ever been produced (and quickly destroyed by hitting the walls of the experiment, as we have no way to store it). The use of antimatter as a power source will have to remain science fiction for the foreseeable future!

In 1942, the Manhattan project was started in the United States with the aim of building a bomb out of fissile material. To that end, an agreement between Winston Churchill and Franklin Roosevelt was signed in 1943, ending the independent British nuclear weapons programme (which was codenamed "Tube Alloys" – clearly, the British have a talent for giving dull names to exciting projects). Early in

the project, Enrico Fermi had demonstrated with an experimental uranium-graphite pile in Chicago that a controlled nuclear chain reaction was possible.

The first nuclear fission bomb was tested in Alamagordo in New Mexico in July 1945. It used plutonium-239 as its fuel. A fission bomb (using highly enriched ^{235}U) was dropped on the Japanese port of Hiroshima on 6th August 1945, and a second (using ^{239}Pu) three days later on Nagasaki. This led to the end of the Second World War.

The dropping of atomic bombs on Hiroshima and Nagasaki certainly represents a major radiation catastrophe with the most severe consequences. It is estimated that 140 000 Japanese citizens were killed in Hiroshima and 80 000 in Nagasaki by the end of 1945. Of those, 110 000 died on the day of the bombings: these deaths were (in almost all cases) due to the explosions and fires, rather than the radiation itself. Since then, thousands more have died from delayed radiation effects and injuries attributed to exposure to radiation released by the bombs. The official number of casualties (up to 2009), according to the Japanese authorities, is 258 000. Many people also suffered from permanent injuries. Due to genetic effects, subsequent generations are also being affected.

In a hydrogen bomb, also known as a fusion bomb or a thermonuclear bomb, most of the explosive energy comes from the fusing of deuterium and tritium. The extreme conditions of temperature and pressure, required for fusion, are provided by the explosion of an atomic bomb. That is, a fusion bomb must contain a fission bomb. The shaping and positioning of the explosives must be extremely precise, in order for the conditions required for fusion to be met. Because of the extremely high power outputs gained from fusion, and the quantity of material that can be fused in

one explosion, hydrogen bombs have higher yields than atomic bombs. The fact that such high power outputs can be gained from fusion is a good indication for future fusion power plants: it is worth noting that as early as the 1950s, fusion reactions were initiated on Earth, and released a huge amount of energy, but not in the way any of us wants to see!

In a hydrogen bomb (see Figure 9.1), the nuclear fission bomb goes off first, and produces large numbers of neutrons and γ rays. The fusion part of the bomb consists in many practical cases of lithium deuteride, a compound of lithium and deuterium (LiD). The neutrons from the fission process react with the lithium deuteride according to the reaction

$$\ {}^{6}_{3}\mathrm{Li} + n \rightarrow {}^{4}_{2}\mathrm{He} + {}^{3}_{1}t \ .$$

The tritium produced in this reaction can fuse with the deuterium from the lithium deuteride. The γ rays from the fission bomb are absorbed by the polystyrene (styrofoam) turning it into a plasma. This plasma tries to expand, which means it exerts an inward pressure on the lithium deuteride, which helps to maintain the conditions so that the reaction can continue. In the fusion process itself,

$$d + t \rightarrow {}^{4}\mathrm{He} + n \ ,$$

further huge numbers of neutrons are produced, which initiate further fission in the surrounding uranium cylinder and in the central uranium rod. At this point, the concentration of high-energy neutrons is so high that even uranium-238 can be made

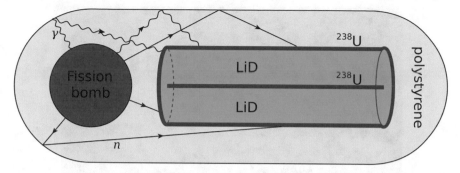

Fig. 9.1 Sketch of a hydrogen bomb. The fission bomb triggers deuterium-tritium fusion in the lithium deuteride (LiD). The fusion then triggers further fissions in depleted uranium (^{235}U). Most of the explosive output is from the fusion. It is important that a sigificant fraction of the nuclear material is consumed before the weapon destroys itself

to fission. This means that, in this case, it is even possible to use depleted uranium. This additional fission creates large quantities of dangerous radioactive fallout.

The first thermonuclear bomb, Ivy Mike, was tested in 1952 by the United States (see Figure 9.2). It vaporised the small island of Enewetak in the Marshall Islands in the North Pacific. The first Soviet thermonuclear bomb was detonated in Kazakhstan in 1953. It was followed by successful tests by the UK (1957), China (1967) and France (1968).

Efforts to negotiate an international agreement to end nuclear tests were initiated by the United States, the United Kingdom, Canada, France, and the Soviet Union. The resulting Partial Test Ban Treaty banned nuclear weapon tests in the atmosphere, in outer space and under water, in 1963. Efforts to negotiate a treaty banning all testing of nuclear weapons (including under ground) have continued, and at the time of writing, a large majority of the world's countries are signed up to the UN's Comprehensive Nuclear Test Ban Treaty, although it can only come into force once still more countries enroll.

As would be expected, there have been further developments to improve the performance of fission and fusion bombs so that they can be used in warfare. One of the examples is the neutron bomb, also called an enhanced radiation weapon. It is a low-yield thermonuclear weapon with relatively low explosive power but a large output of neutrons. Because of the strong biological effect of neutrons (and therefore their high radiation weighting factor), these bombs can cause massive human casualities while leaving most of the infrastructure intact. They also create relatively little radioactive fallout. Many attempts have also been made to minimise the size of these weapons, but this turned out to be difficult, because a certain critical amount of fissile material is needed to make a bomb work.

The size of explosions caused by nuclear weapons is measured using a slightly odd unit, namely the ton of TNT (or kiloton for a thousand tons, or Megaton for a million tons). It refers to the amount of TNT (trinitrotoluene, a common explosive) which would be needed to produce an explosion of that size. One kiloton of TNT

Fig. 9.2 The mushroom cloud from the Ivy Mike test on Enewetak, an atoll in the Pacific Ocean on 1st November, 1952. (Image credit: The Official CTBTO Photostream – "Ivy Mike" atmospheric nuclear test, November 1952; (CTBTO: Comprehensive Nuclear-Test-Ban Treaty Organization) www.flickr.com/photos/ctbto/6476282811)

is an explosive output of about 4.2 trillion Joules (4.2 Tera-Joules). The use of this unit makes it clear how powerful nuclear weapons are, because clearly, collecting thousands of tons of TNT in one place and bringing it to explosion would not be practical.

The destructive power of nuclear weapons is caused by the pressure wave (50 %), heat emission (35 %), and γ- and X-ray radiation (15 %). A 1-Megaton weapon creates a crater of about 2 km across in, typically, 10 seconds. The largest weapon ever produced was the Tsar Bomba (about 55 Megatons), which was tested by the Soviet Union in 1961. It was capable of producing a crater 10 km across. The explosive power of Tsar Bomba was 1500 times that of the weapons dropped on Hiroshima and Nagasaki put together.

Quite apart from the risk of total nuclear armageddon, and from the direct harm to victims in Hiroshima, Nagasaki, and some test sites, nuclear weapons have had damaging effects on the population. Nuclear weapons tests, particularly the atmospheric tests in the early 1960s, released large quantities of radioactive material into the biosphere: considerably more than the Chernobyl disaster, although Chernobyl has gained much more international attention. Fortunately, due to the decay of most of the radioactive substances involved, the doses now being received from this fallout are less than 1 % of the natural background radiation.

In addition to the nuclear weapons mentioned, it is important to consider the possibilities of other nuclear threats. The major terrorist threat in this context (assuming they never got their hands on a nuclear weapon) comes from *dirty bombs*. These are explosive devices which use a conventional weapon (for example a simple dynamite bomb) to disperse radioactive material. The purpose of scattering the radioactive material is to contaminate a large area, but it will not cause mass casualities from radiation poisoning. The main purpose is to spread panic and fear of radiation. Fortunately, there has never been a dirty bomb attack, but it might be possible for a terrorist to cause a large amount of terror, out of proportion to the real danger, by including a small, relatively easily obtained radioactive source, such as one stolen from a hospital or from an industrial plant. Clearly, weapons involving radioactive material provide an impressive list of potential horrors!

Summary

Nuclear weapons were developed during and after the Second World War. They come in two distinct types: fission bombs and fusion bombs. Fusion bombs have higher yields, and must include a fission bomb as a trigger. Fortunately, nuclear weapons testing is currently banned, a ban observed by almost all nations on Earth. As a separate possibility, terrorist attacks using radioactive material are possible.

Chapter 10
What Happens When It Goes Wrong?

Till now man has been up against Nature; from now on he will be up against his own nature.

Dennis Gabor 1900–1979

Many radiation accidents in the fields of medicine and technology are caused by loss and careless disposal of radioactive material. The reason for unnecessary exposures is frequently improper storage of disused radioactive sources. Discarded sources have been found in scrapyards by children, who were pleased to find some pieces of (for example) nice-looking silver-coloured cobalt metal: tragically, this was highly radioactive and dangerous. Table 10.1 shows a number of examples of losses of radioactive sources and events of accidental irradiations.

Radiation accidents in large manufacturing plants and nuclear-medical sections of hospitals happen most frequently because the most basic safety rules are not in place, or are ignored. It is also essential that the maintenance personnel are suitably

Table 10.1 Radiation accidents with radioactive sources

Year	Country	Source	Application	Activity	Deaths	Comment
1962	Mexico	^{60}Co	Inspection of metal	185 GBq	4	Discovered by children
1963	China	^{60}Co	Seed irradiation	370 GBq	2	Discovered by children
1978	Algeria	^{192}Ir	Industrial radiography	925 GBq	1	Accidental exposure of workers
1987	Brazil	^{137}Cs	Nuclear medicine	370 GBq	4	Discovered by children
2013	Mexico	^{60}Co	Radiotherapy	111 TBq	Not yet known	Stolen in transit: fatal for thieves?

© Springer International Publishing Switzerland 2016
C. Grupen and M. Rodgers, *Radioactivity and Radiation*,
DOI 10.1007/978-3-319-42330-2_10

© by Nick Downes

trained and are aware of the radiation risks. Radiation-protection regulations must be meticulously respected, otherwise accidental irradiations may occur.

An example of gross human failure is the radiation accident in 1992 in a hospital in Indiana, Pennsylvania (USA), in the context of brachytherapy. Brachytherapy is the treatment of cancer by the placing of a radiation source into the immediate vicinity of the tumour, either inside the body of the patient or on the skin. The source can be inserted into a suitable orifice of the body (such as the colon) or into cavities created artificially by surgery. In the Indiana hospital, an elderly lady was irradiated with an iridium source, but the staff forgot to remove the source after the treatment. When the patient excreted the catheter containing the source four days later, the catheter, including the source, was thrown into the garbage by a nurse. The patient died the following day without realising that the death could have been related to the excess radiation. During the time of storage and garbage collection, 90 days in total, many people were accidentally exposed to radiation. The loss of the iridium source was only discovered when radiation monitors in the waste-management facility triggered an alarm.

Only after the end of the Cold War has it become public how many radiation accidents occurred in military operations in the past. These include plane crashes with nuclear weapons on board, nuclear-operated submarines being sunk, and the loss of missiles and satellites which were carrying radioactive materials.

Nuclear reactors gain energy by fission (see Chapter 8). During the operation of these reactors, 'incidents' of different severities occasionally occur. The most severe accident in a nuclear reactor to date happened at a boiling-water reactor in Chernobyl, modern-day Ukraine, in 1986. This catastrophe was the result of a flawed and poorly-executed experiment at an inherently unsafe reactor. In this experiment, a large fraction (perhaps 50 %) of the radioactive content of the plant was released into the environment. The water-cooled graphite-moderated reactor contained a total of 150 tons of slightly enriched uranium (2 % ^{235}U), and its radioactive fission products, with a total activity of 7×10^{19} Bq. The gaseous components (radioactive ^{85}Kr $(3.3 \times 10^{16}$ Bq) and ^{133}Xe $(6.5 \times 10^{18}$ Bq)) escaped completely from the reactor. The solid fission products (iodine-131, caesium-137, strontium-90, and others) were also released in large quantities, in the form of dust. Table 10.2 shows the composition of the plume of radioactive material released by the Chernobyl explosion.

Since then, the short-lived fission products have decayed, so that the remaining sources of radiation are the longer-lived isotopes ^{137}Cs and ^{90}Sr. These two nuclides will remain in the biosphere for many decades.

Figure 10.1 shows the average ^{137}Cs content of human bodies in Western Europe over the period from 1960 to 1995. For these people, the (radioactive) caesium exposure due to the reactor accident in Chernobyl was nearly as high as the caesium exposure from all the above-ground nuclear weapons tests in the atmosphere in the 1960s. These surface tests released about 3 tons of plutonium which was distributed worldwide, predominantly in the northern hemisphere. In addition, the neutrons released produced large quantities of ^{14}C and tritium (^{3}H) in the atmosphere.

The passage of the Chernobyl plume over Europe led to distinct fallout patterns for different radioisotopes. Exposures to ^{131}I were caused mainly by drinking milk from cows that had eaten fallout-tainted vegetation. Other exposures arose from breathing contaminated air or eating contaminated food: some mushrooms concentrated the

Table 10.2 Estimated composition of the Chernobyl plume

Isotope	Fraction (%)	Half-life
Xenon-133	55.1	5 days
Iodine-131	14.9	8 days
Tellurium-132	9.7	3 days
Iodine-133	7.7	21 hours
Neptunium-239	3.4	2 days
Barium-140	2.0	13 days
Ruthenium-103	1.4	40 days
Strontium-89	0.9	52 days
Caesium-137	0.7	30 years
Zirconium-95	0.7	65 days
Cerium-141	0.7	33 days
Strontium-90	0.07	28 years

Fig. 10.1 Average ^{137}Cs content of humans in Western Europe from 1960 to 1995. For regions further from Europe, the Chernobyl caesium peak is much less pronounced. The first peak is similar throughout the whole northern hemisphere

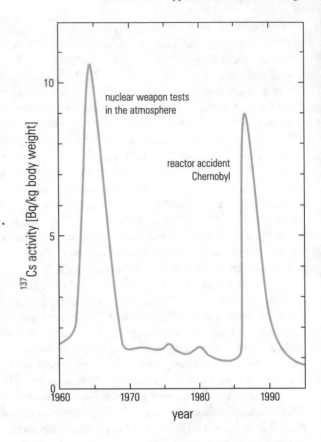

caesium-137 in their tissues, and anything that ate them (including deer, reindeer and humans) absorbed that radioisotope. The contamination of the ground showed large variations, depending on the weather conditions, particularly rainfall and wind. It was possible to detect the Chernobyl plume elsewhere in the northern hemisphere, in particular in North America and Japan, but outside Europe and Russian Asia, the exposures were very low.

In Western Europe, exposures due to Chernobyl were not particularly high, and were (apart from a few exceptional cases) well below any safe dose limits used in radiation protection, and also well below a typical annual dose from background radiation. Estimates of the average additional exposure of the population in Western Europe due to the Chernobyl accident are under 0.5 mSv (this is the figure for 1986: the exposures for subsequent years were too low to be significant). Estimating the number of additional cases of cancer for this exposure is difficult. There is a standard estimate of risk (see Section 7.5), which is 5 %/Sv, and is likely to be a large overestimate for this case. The same number can also be written as one cancer case per 40 000 people per 0.5 mSv exposure. This suggests a number of instances of cancer, across 400 million people, of 10 000. This number can be regarded as a maximum possible limit, and even this many cases would be lost in

the 'noise' of other cancer cases in such a large population over a period of several decades (in individual cases it is not possible to establish correlations between cancers induced by this radiation and 'normal' cancers).[1]

The situation within a few hundred kilometers of Chernobyl is completely different. The rate of cancer of the thyroid gland among children has significantly increased. The leukaemia rates, however, showed no correlation with the Chernobyl disaster. In fact, they stayed constant. Large numbers of people living in Belarus (2 million) and the Ukraine (1 million) received estimated doses around 20 mSv. If the latest risk factors are taken at face value, this would correspond to an additional cancer rate of 3000 cases. Another 1000 cases are expected among the rescue workers and people staying in the immediate vicinity of the reactor. Because of the long latency period for these cancers, the peak in the number of cases should occur in the next few years.

These frighteningly high expectation values are based on the most recent estimates of the risk factors (5 %/Sv). The number of cases of cancer actually observed in the former Soviet Union, and in particular the non-observation of an increase in the leukaemia rate, appears to indicate that the risk factors used somewhat overestimate the real risk. The official number of deaths is given by the authorities as 31. The

---

[1] Because of the uncertainty of health effects at low doses, the ICRP writes in their recommendations in 2007 that, at least for the purposes of health planning, "it is not appropriate … to calculate the hypothetical number of cases of cancer or heritable disease that might be associated with very small radiation doses received by large numbers of people over very long periods of time." The effect of low doses is influenced by the elaborate defence system of the human body, which allows the repair of minor radiation damage. There is even some evidence that the human immune system is activated by low-level radiation (a process called hormesis): this would act to reduce or eliminate the effect of the radiation, or might even provide a small health benefit.

International Atomic Energy Agency has independently estimated the total number of deaths (including past and future deaths) as a consequence of this nuclear disaster to be 4000.

The radiation accident in the recycling plant Tokaimura in Japan on 30th September 1999 represents an example of a criticality accident which could have been avoided. Three of the plant's workers had filled a large tank with a highly enriched solution of uranium using buckets. The radiation-protection regulations would certainly have forbidden the carrying of uranium solution in buckets in this way. Most nuclear power plants (boiling-water and pressurised-water reactors) normally use uranium which is enriched to a level of 3 % to 5 % with ^{235}U. Because Tokaimura was a plutonium production project, the level of enrichment in this case was 18.8 %. The volume of highly-enriched solution poured into the tank was a large quantity in the context of nuclear power, at 16 kg (6–7 litres): this is several times the minimal amount for a self-sustaining fission reaction. To handle such a large amount in a single container was forbidden under all circumstances. Under these conditions, a chain reaction will start instantaneously and explosively. The workers reported that they observed a blue flash when the solution had been poured in. It is safe to conclude that this was the characteristic blue light called Cherenkov radiation – this is light produced by fast-moving electrons in water. This was a clear indication that a self-sustaining chain reaction was taking place: the fast electrons originated from the decay of the neutrons produced by the process. The critical state of the uranium solution persisted for 18 hours. The fission conditions were maintained over such a long period because the container was surrounded by water, whose initial purpose was to cool the solution. Water acts as a good moderator and neutron reflector, so unfortunately, it was the coolant which allowed the chain reaction to continue!

One of the workers received a dose of 17 Sieverts. In spite of intensive medical care (e.g. blood transfusions) he died after 83 days. The other two workers received doses of at least 8 and at least 3 Sieverts. Other workers on the site of the recycling plant were mainly exposed to penetrating neutron radiation resulting in multiple individual doses of more than 20 mSv. In total several grams of ^{235}U were fissioned. This chain reaction generated fission products corresponding to activities of 10^{16} to 10^{17} Bq, and of these products, about 1 % were released into the environment. In total, this accident resulted in two deaths.

In this example it became evident that the safety regulations were grossly violated. No container, with the exception of the operational reaction vessel, should contain an overcritical amount of fissile material, so to fill a single container with several times this amount would be unthinkable in a well-run plant. Even worse, this container was immersed in cooling water, which acted as moderator and neutron reflector. After the event it was observed that the safety regulations in this recycling plant were habitually disregarded. Also the handling of the incident indicated that emergency procedures and evacuation plans either did not exist or were ignored.

The more recent nuclear disaster in the Fukushima Daiichi nuclear power plant in March 2011 resulted in a meltdown of three of the plant's six reactors. The failure was initiated when the plant was hit by a giant tsunami triggered by an earthquake.

The result was an enormous radioactive discharge, whose consequences were only less serious than those of Chernobyl because a large majority of the fallout fell on the ocean. The boiling water reactors were flooded by the tsunami causing a prolonged loss of power for the emergency systems. In particular, the cooling systems broke down causing the fuel rods in the reactor to heat up, leading to a meltdown. The chemical interaction of the hot zirconium cladding of the fuel rods with water created large amounts of hydrogen gas, which is highly flammable. The resulting hydrogen-air explosions destroyed large parts of the reactor buildings, and as a consequence, repeated spills of contaminated water into the sea have occurred, and are still going on even several years after the meltdown. Two people were taken to hospital with radiation injuries (although this must be seen in context: about 20 000 people were killed due to the earthquake and tsunami, and the subsequent evacuation). There is no known way to decontaminate the affected forests and land around the reactors from the main isotopes (^{137}Cs, ^{134}Cs, and ^{90}Sr). The cancer risk due to late radiation effects was, however, estimated to be small for most parts of the surrounding area. Nonetheless, it was possible to observe an increase in thyroid cancer, especially among young people. It is important to note that the psychological consequences of such a disaster and the subsequent evacuations and relocations are enormous. Indeed, it appears that several elderly people were killed by the effects of the evacuation process itself (because they were too frail for it) – this is despite the fact that, given their ages, their cancer risk from the radiation was low, so they might well have survived if they had stayed in the contaminated area.

The disaster was caused because the plant was essentially unprotected against a tsunami 14 metres high (the plant's seawall was only 10 metres high). In addition, the Tokyo Electric Power Company (TEPCO) had no real plan to cope with such a nuclear emergency. Certainly, the backup power supplies, which should keep the cooling running and prevent meltdown even after a large mechanical failure, were insufficient. Some risks of operating nuclear power plants in areas that are frequently affected by earthquakes seem to have been downplayed, compromising operational safety. This can be compared with the situation in Chernobyl where an experiment was being performed to check whether the running of turbines after a shutdown would still provide enough power for the emergency systems. To find out, all automatic safety systems were shut down on the Chernobyl site, and the reactor was operated manually.

Both the Chernobyl and Fukushima disasters were essentially man-made, and the direct causes were in principle forseeable. It must be pointed out again and again that appropriate safety rules must be followed to prevent such disasters, or at least to be adequately prepared for any possible nuclear emergencies if they do occur.

From the incidents mentioned above, some general lessons on the causes and management of radiation accidents can be learnt. Frequently it takes too long for an accident to be identified. It might be suspected that there is even a substantial number of incidents that the public will never find out about. Time wasted can have serious implications: a delay in handling the accident normally leads to a deterioration of the situation. Prevention of accidents is essential. The appropriate safety regulations must be known to the relevant people and they must be respected. It is essential

that all members of staff have appropriate training. It goes without saying that safety systems must not be switched off or circumvented (as they were at Chernobyl). These safety systems must have uninterruptable, independent power supplies (as they did not at Fukushima). Common sense helps in most situations. Special attention must be paid to the human factor: it happens again and again that safety regulations are ignored.

The responsibilities of the personnel have to be clearly defined, and the authorities must be able to recognise the scale of an accident. Therefore it is essential to have emergency plans prepared in advance. These plans must cover multiple levels: the reactor itself; the non-nuclear components (such as turbines and transformers); and the process for informing the public about any measures they must take (such as evacuation, or the taking of iodine tablets). It is also important that these emergency plans are well-known to staff, and rehearsed in drills, before critical situations or accidents arise.

In the following section, some examples of smaller-scale irradiation will be discussed. The aftermath of the dropping of nuclear bombs on Japan in 1945 is discussed in Chapter 9.

10.1   Smaller-Scale Radiation Incidents

In many cases, the loss of radioactive sources and accidental irradiations have been recognised only after a long delay In the first example of Table 10.1, nearly all the members of a family were killed. In March 1962 in Mexico, a boy found an abandoned, pencil-sized radiography gauge, containing a highly radioactive 185 GBq ^{60}Co source, in a scrapyard. The source was in principle sealed, but it was broken. The boy played with the source and took it home. The boy's mother then found the source and placed it on the kitchen shelf. The 10-year-old boy died in April and his mother followed in July. Nobody suspected that the deaths were related to ionising radiation from the source. The common cause of the deaths was only recognised when a 3-year-old child in the family also died. Another casualty followed in October. Only the father of the family survived, because he spent relatively little time in the home.

Radiation accidents in the military sector have often been kept secret, especially during the Cold War. Typical examples originate from the disastrous management of military installations and the loss of nuclear weapons by the superpowers, the United States and the former Soviet Union. Air accidents led to the loss of quite a large number of nuclear weapons at sea. Also, failed nuclear-missile launches released substantial amounts of radioactive material. In particular, nuclear-powered submarines that have sunk have discharged large quantities of nuclear material into the oceans. In addition, the atmosphere has been polluted by radioisotope batteries (see Section 4.6) on board satellites, which broke up when re-entering the atmosphere, and released their plutonium as dust. The potential dangers arising from secret military operations became obvious when a Soviet nuclear submarine was wrecked in the Atlantic in 1961. The Soviet authorities were anxious that the submarine might be recovered by other powers and decided not to abandon it, but instead gave the order to repair it on the spot. The authorities accepted that several members of the crew would receive rather high doses during the repair operations. At least eight of them died as a result of high exposure.

In the 1940s radium-beryllium sources (see Section 4.3) were fabricated manually in laboratories. A strong one-gram Ra-Be source can be very dangerous if it allows any leakage. Radium decays into the radioactive noble gas radon which can easily be released from the source. If the source was not tightly sealed in the manufacturing process, it was almost impossible to handle it after a few hours because of the continuous escape of radon. A substantial radiation exposure must be accepted to seal a leaking source afterwards. During the fabrication and sealing of such a source, co-workers of Enrico Fermi (a physicist who won a Nobel Prize) received a γ dose of about 2 Sv. This exposure halved their number of white blood cells (leucocytes), which recovered only slowly back to normal. However, late radiation effects were not observed.

A special case of *nuclear terrorism* was the murder of Alexander Litvinenko in 2006. Alexander Litvinenko was a former officer of the Russian Committee for State Security (KGB) who fled from prosecution in Russia and received political asylum in the UK. In the UK, he supported a media campaign against the Russian government and worked for the British security services. In November 2006 he suddenly fell ill, showing the typical symptoms of radiation sickness, which were not immediately recognised as such. Earlier that day he had met two former KGB officers who are suspected of poisoning him with polonium-210, a colourless and odourless substance, which they had slipped into his tea. ^{210}Po is an α-emitter decaying to lead-206 with a half-life of 140 days. It appears that at least 50 μg of ^{210}Po was used (corresponding to about 10 GBq), and that Litvinenko ingested about 10 μg: this is approximately 100 times the lethal dose of this isotope. The amount of polonium used would have a commercial market value of tens of millions of dollars! With relatively little shielding, a polonium source can be carried through airport screening devices without setting off any alarms, because the α rays have very short range. Due to the high biological effectiveness of α rays, this isotope is extremely dangerous when incorporated. Because of the suspected involvement of the Russian government, Litvinenko's death led to serious diplomatic difficulties between the UK and Russia.

Summary

Radiation accidents are frequently caused by the ignoring of elementary safety rules. The loss of radioactive sources or careless disposal of radioactive waste can also lead to unnecessary exposures. Many radiation accidents in military installations or operations have been kept secret. The largest accident to date occurred in the then Soviet nuclear reactor in Chernobyl in 1986. This catastrophe was caused by a fatal experiment on the reactor, and the absence of a reliable safeguard system. The nuclear reactor, during meltdown, produced a radioactive cloud that floated over the whole northern hemisphere.

Chapter 11
What About Non-Ionising Radiation?

*High frequency radiation is sometimes thought to be the cause
of cancer, while low frequency radiation is generally assumed to
be harmless.*

Susan Dean 1945– and Barbara Illowsky 1959–

Most of this book has concerned ionising radiation: chiefly α, β, and γ rays and
neutrons, but also X rays. γ radiation and X rays are electromagnetic waves: they
differ from visible light or microwave radiation only by their energy (or, equivalently, by their frequency or wavelength). It is only natural to ask to what extent
electromagnetic radiation of other frequencies might be dangerous for humans.

For electromagnetic radiation, energy and frequency are directly related, and
wavelength is inversely related to the other two. This means that a discussion about
high-energy radiation could equivalently be about high-frequency radiation, or short-
wavelength radiation. Similarly, low-energy, low-frequency and long-wavelength
radiation are the same.

Depending on the energy, electromagnetic radiation has very different effects.
Table 11.1 and Figure 11.1 give some characteristics of frequency ranges along with
their commonly used names and typical applications.

The transitions between the different characteristic ranges are not particularly
well-defined. To ionise atoms in human tissue, an energy of approximately
30 eV is required. Therefore, radiation with frequencies below 10^{16} Hz (corresponding to 30 eV) is termed non-ionising. Rather than from ionisation, the potential for
harmful consequences from non-ionising radiation comes from two other effects:
heating and electric currents.

The higher-frequency forms of non-ionising radiation (a large range of wavelengths between UV and microwaves) cause significant heating of the human body
if they have high enough intensity. This means that the most relevant quantity for
assessing potential harm is the radiation power delivered, per unit mass of tissue.

© Springer International Publishing Switzerland 2016
C. Grupen and M. Rodgers, *Radioactivity and Radiation*,
DOI 10.1007/978-3-319-42330-2_11

Table 11.1 Different types of electromagnetic radiation, with a typical frequency, wavelength and energy for each one

Classification	Frequency (Hz)	Wavelength (m)	Energy (eV)
Mains, AC current	50	6×10^6	2×10^{-13}
Long-wave radio	6×10^4	5000	2.5×10^{-10}
Short-wave radio	3×10^6	100	1×10^{-8}
Ultrahigh-frequency radio	10^8	3	4×10^{-7}
Mobile-phone communication	3×10^9	0.10	1×10^{-5}
Microwaves	6×10^9	0.05	2×10^{-5}
Radar	3×10^{10}	0.01	1×10^{-4}
Infrared	10^{12}	3×10^{-4}	4×10^{-3}
Visible light	6×10^{14}	5×10^{-7}	3
UV	1.5×10^{15}	2×10^{-7}	6
X rays	10^{19}	3×10^{-11}	4×10^4
γ rays	over 2×10^{20}	under 1×10^{-12}	over 10^6

For a whole-body exposure, a limit of about 0.1 W/kg is recommended.[1] This limit is a few percent of the exposure a person would get from exposing their full body to sunlight in the day at temperate latitudes. For partial-body doses, higher limits are tolerated (head and body 2 W/kg, legs and arms 4 W/kg).

For the lowest-frequency forms of radiation, which means all forms of radio waves below microwaves, it is very difficult to produce a significant heating effect. The only significant source of potential harm is the currents which are induced in the human body by these waves, as they can disrupt the functioning of nerve and muscle cells, if they are strong enough.

Electromagnetic radiation is made up of an electric field and a magnetic field. Electric field strength is measured in Volts per metre (V/m), and magnetic field strength in Tesla (T).[2] Either of these components can induce a current in the tissues of the body. To assess the impact of the radiation on the tissue, the intensity of a current (not just its total value) must be considered, so the best quantity to use when considering potential biological effect is the current density, measured in Ampères per cm^2.

There are already natural currents in the human body for conduction in the nervous system. It is these currents which make it possible to measure the activity of the heart with an electrocardiogram (ECG), and that of the brain with an electroencephalo-gram (EEG). The associated natural current densities vary between 0.1 $\mu A/cm^2$ and 1 $\mu A/cm^2$. In setting limits on the permissible amount of low-frequency radiation humans should be exposed to, it is important to take into account that the human body is tolerant to at least this level of current density.

---

[1] By the International Commission on Non-Ionizing Radiation Protection (ICNIRP).

[2] Technically, this is the magnetic flux density, but the distinction between these quantities is subtle, and is not preserved in this book.

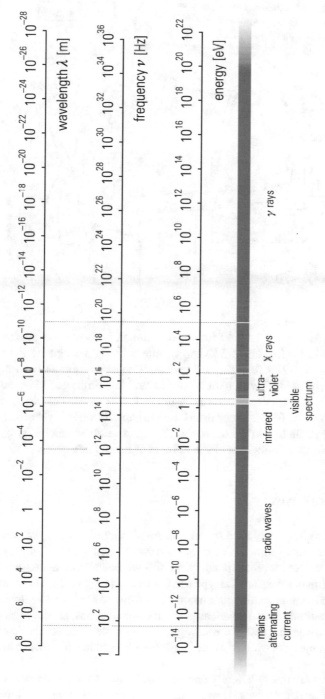

Fig. 11.1 Spectrum of electromagnetic radiation. Although it is not a range, just a point within radio frequencies, mains AC is provided for reference

A current density of $0.1\,\mu\text{A}/\text{cm}^2$ in the human body can be produced by an external electric field of about $5\,\text{kV/m}$, or alternatively a magnetic field of about $100\,\mu\text{T}$ (assuming $50\,\text{Hz}$ frequency). This magnetic field is similar in strength to the Earth's magnetic field. Therefore, it is only natural that limits of $5\,\text{kV/m}$ and $100\,\mu\text{T}$ have been proposed for continuous exposure to low-frequency electromagnetic AC (alternating current) fields. These limits are perhaps 20 times higher than the fields experienced by standing in the open, directly underneath an electricity pylon (note that being inside a building would shield most of the pylon's field).

11.1   Microwaves

Microwaves are frequently used for telecommunication. They are generated by specialised vacuum tubes in which electrons are accelerated and deflected by magnetic fields. These vacuum tubes, depending on design, are called magnetrons, klystrons or gyrotrons (magnetrons are the type used in microwave ovens). These devices can produce a more or less continuous stream of microwave photons, or they can also be operated in a pulsed mode. As well as in telecommunications, microwaves have been used in radioastronomy and in cosmology: it is possible to detect microwaves from astronomical objects or even from the Big Bang (the cosmic microwave background radiation).

Microwaves are not sufficiently energetic to ionise atoms. It has not been shown conclusively that microwaves at low intensity have adverse biological effects (indeed, low intensity microwave radiation is produced by all warm objects, including the

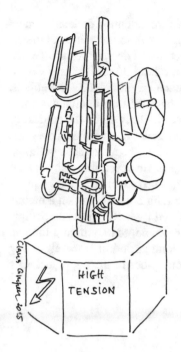

human body). However, injuries will occur on exposure to high microwave intensities due to the heating of the body absorbing the radiation. Especially vulnerable are the lens of the eye and the cornea, since they do not contain blood vessels which can carry away the heat. On irradiation, the lens of the eye may become opaque, causing a cataract.

The penetration of microwaves depends on the frequency. Microwaves can heat a thin layer of the surface of the skin to intolerable temperatures, which means they can be used for non-lethal weaponry. Also, pulsed microwaves of GHz frequencies, if they exceed peak powers of $100 \, \text{mW}/\text{cm}^2$, will produce heating of the brain, causing pressure waves to travel through the skull. These waves can cause audible clicks, generated inside the head without the need for any receiving device. These clicks might be disruptive to the sense of hearing, could have psychologically destructive effects and can cause dizziness or headaches. Microwave ovens with poor sealing can be hazardous because of the heating of the skin, but the intensity is not high enough to cause significant pressure waves. Another possible danger comes from the chemicals used in the microwave generators. They often contain beryllium oxide insulators: if these are crushed and the dust inhaled, it can lead to incurable lung disease. It is because of this that they are usually marked "do not disassemble".

Currently it is a matter of debate whether it makes sense to limit mobile-phone[3] radiation solely because of its heating effect on body tissues. There have been

---

[3]This book uses the term "mobile phone". "Cell phone" is the term common in some English-speaking territories.

unconfirmed reports of headaches, sleep disturbances and attention deficit disorders being caused even by low-level microwave radiation, down to intensities as low as 1 mW/cm^2. One suggestion for a non-thermal effect has to do with resonance: the idea that the wavelength of the microwave radiation (about 30 cm) corresponds to the diameter of the head, and that this correspondence might magnify the effects. It appears that pulsed radiation is potentially more harmful than continuous wave radiation. This certainly requires further research, particularly given the context of rising levels of low-frequency radiation from mobile phones and wireless networks. The claim has been made that there is no safe threshold for electromagnetic pollution (a claim also sometimes made about ionising radiation). However, this claim is hard to believe given the natural amounts of background (microwave) radiation.

It is often said that people get a warm ear after a long conversation on a mobile phone, and that this has to do with the radiation emitted by it. This is a compelling, but incorrect, argument. People do get a warm ear whenever they hold a thermal insulator, such as a phone, against their head for a long period: it is merely acting like an ear muff. The ear would also get warm if any other insulating object, or a turned-off mobile phone, was used.

"I no longer suffer from cold ears, because I use my mobile phone in winter!"

© by Claus Grupen

Table 11.2 EU proposed maximum allowed power per unit area for various radiation sources in the microwave frequency range

Radiation source	Frequency range	Maximum allowed power per square metre
Ultrashort wave	88–108 MHz	$2\,W/m^2$
VHF	174–216 MHz	$2\,W/m^2$
UHF	470–890 MHz	2–$4\,W/m^2$
D network	890–960 MHz	$4.5\,W/m^2$
E network	1710–1880 MHz	$9\,W/m^2$
UMTS	around 2 GHz	$10\,W/m^2$
Microwave ovens	around 2.5 GHz	$10\,W/m^2$

In typical mobile-phone communications, peak powers of up to 2 Watts are reached. The limits presently proposed in radiation-protection regulations for non-ionising radiation depend on the assumption that the biological effect of microwave radiation is purely due to its heating effect on tissue. This means that the limits are expressed in terms of the heat power per unit area. As an example, some limits which are recommended in the European Union are compiled in Table 11.2.

For comparison: the radiation power of the Sun at the edge of the atmosphere is about $1400\,W/m^2$. This radiation power is reduced as it passes through the atmosphere, and to different extents in different weather conditions. Under a cloudy sky, about $100\,W/m^2$ can be measured at sea level at intermediate latitudes (like those of Europe and the USA).

It is worth mentioning that there are medical procedures which benefit from the use of microwaves. In microwave diathermy, radar waves of frequencies around 500 MHz are used. It is commonly applied for muscle relaxation since it has a heating effect, in particular on the deeper tissues of the body, and provides short-term pain relief.

11.2 Ultraviolet Light

In the ultraviolet range, the heating of the skin becomes particularly important. In clear weather conditions, the solar radiation power on the ground at temperate latitudes can be $300\,W/m^2$ (about $15\,W/m^2$ in the UV), leading typically to a reddening of the skin in less than half an hour, for people with sensitive skin. This is due to the fact that the radiation is absorbed at low skin depths and transformed into heat.

Since UV radiation can cause skin cancer, we will deal with this type of radiation in a little more detail. The ultraviolet spectrum is subdivided into four ranges: UVA, UVB, UVC and extreme UV. Extreme UV, the highest energy form, is completely absorbed by the Earth's atmosphere and need not concern us here. The lower three ranges are shown in Table 11.3.

The main effect on humans of absorption of low-frequency (or equivalently low-energy) UV radiation, is heating of the body's surface tissues. UVA radiation causes

Table 11.3 The ranges of UV radiation. Higher frequencies are absorbed more strongly by the atmosphere, so percentages of total solar energy are given for the top of the atmosphere (external) and for sea level

Type	Wavelength range (nm)	Energy range (eV)	Energy Percentage	
			external	sea level
UVA	400–315	3.1–3.9	6.7%	5.7%
UVB	315–280	3.9–4.4	1.3%	0.3%
UVC	280–100	4.4–12.4	under 0.2%	insignificant

tanning of the skin, with the maximum effect being obtained at a wavelength of 340 nm. The threshold value for pigmentation without burning is around 10^5 W s/m^2 (for example, 10 W/m^2 for about three hours). If the dose of UVA is too high, sunburn results. Sunburn is not only created by the thermal effect of radiation. UV radiation also induces chemical reactions in the skin which for high intensities can lead to the release of cellular poison. These effects are stronger for UV of higher energy, which is why the daily safe limit for UVC is considerably lower at 30 W s/m^2: it is fortunate that natural UVC is practically absent at the Earth's surface. UVC radiation is sufficiently energetic to move electrons in the skin's molecules to higher energy states (excitation). For the highest energies, ionisation processes can even occur. The ionisation energy required depends on the type of atom. Most of the biologically relevant atoms like carbon, oxygen and nitrogen are only ionised for wavelengths below 100 nm, i.e. energies higher than UVC. As well as ionisation, the breaking of molecular bonds can occur. Molecules produced after the breaking of bonds (free radicals) tend to interact more intensively with the tissue leading to possible biological hazards.

"Better safe than sorry!"

In historical times this effect was used in bleaching. Ultraviolet radiation from the Sun produces small amounts of hydrogen peroxide from water using oxygen from the air. The peroxide can destroy colourants embedded in cloth, and thereby whiten clothes. Also, the exposure of plastic materials to UV radiation frequently causes brittleness.

Apart from tanning and sunburn, also inflammations of the cornea (keratitis) and cataract of the eye may occur. A daily safe limit to avoid these effects on the eye would be about $50\,000\,\mathrm{W\,s/m^2}$ (for example, $10\,\mathrm{W/m^2}$ hitting the eye for one and a half hours). It is hard to reach this limit without prolonged periods looking directly at the Sun, or strong, artificial UV sources. Nonetheless, it is still a bad idea to look directly at the Sun for any period of time, no matter how short.

There are also so-called late effects, i.e. effects that may occur after a long latency period. The probability of these effects depends on the total radiation dose (summed over time). The most serious late effect is skin cancer, of which there are two important types. The first type is basal cell carcinoma. This type of cancer, the risk of which rises with exposure to UVB, is the most common and least lethal form of all cancers. Early treatment is important, and is usually effective. In contrast, high-frequency UV radiation favours the formation of malignant melanoma, a very severe medical condition. In this case, early detection of the skin cancer is essential, since malignant melanoma has the tendency to form metastases (i.e. to spread). In most countries with intense sunshine, skin cancer is the cancer type with the most rapidly rising number of cases.

The negative effect of UV, particularly UVC, on biology can also be useful. With intense UVC radiation, microorganisms, and especially bacteria, can be killed. Therefore, this type of radiation is suited for disinfection, particularly of transparent media like water and air.

11.3 Lasers

Lasers have specific radiation-protection regulations. The biological effect of laser radiation is related to its extremely high power density, i.e. most lasers are dangerous not because they put out a high power, but because that power is concentrated on a small area. Laser beams are frequently used for alignment and laser-optical experiments. With laser beams, very small holes can be burnt, tissue and material can be cut, and there is even active research into using lasers to induce nuclear fusion (see Section 8.2.1). Lasers are also applied in surgery, for example to remove very thin layers of tissue to correct errors in the lens of the eye.

The proposed limits for lasers are rather complex, but perhaps the most crucial aspect is the sensitivity of the human eye. A commercial laser pointer emits mostly in the wavelength range 630–680 nm (red). It has an output power of less than 1 mW and a beam cross section of about 2 mm by 2 mm. Even this leads to radiation power densities around $200\,\mathrm{W/m^2}$, comparable to looking directly at the

Sun in the late afternoon. In addition to red lasers, green and blue lasers are now available (typical wavelengths 532 nm and 447 nm respectively), which have better visibility due to their higher power, which also means they carry an increased risk to the eyes.

In the handling of laser beams, one must be aware of reflecting surfaces, as even reflections can be harmful. Particular care has to be taken for lasers in the pulsed mode. At present there are femtosecond lasers which produce high-power laser pulses of astonishingly short duration (10^{-15} seconds). Because of the large number of applications and laser variants, lasers are subdivided into different classes according to their potential hazard:

Class 1:   Very safe lasers. Lasers fall into this class if it is not possible (even with magnifying equipment of moderate strength) to exceed any safety limits using them.

Class 2:   Lasers of low radiation power in the optical range. Although shining these at people's eyes is inadvisable, the blink reaction is fast enough to prevent permanent damage. For continuous-wave, visible lasers the power limit is 1 mW. 'Red' laser pointers belong to this class.

Class 3:   Optical lasers with powers over 1 mW. A direct view into the laser beam is dangerous, as is a direct reflection, such as from a mirror. Diffuse reflections (for example from a rough surface or from paper) of unfocused lasers of this class are not dangerous. Typical laser powers of this class range from 1 to 5 mW.

Class 4:   High-power lasers for which diffuse reflections are also hazardous. From these lasers, there is a risk of injury to the skin, as well as to the eyes. They can also be strong enough to cause ignition of flammable material. This is the top level of power, and these devices require maximum care in handling.

The laser powers noted above refer to continuous-wave lasers operating in the visible light range. Infra-red lasers have higher limits on their power, and the structure of limits for pulsed lasers is more intricate. The details of the limits on pulsed lasers for different frequencies are published by the International Commission on Non-Ionizing Radiation Protection (ICNIRP) in their guidelines.

Scientists and technicians handling lasers have to wear safety goggles which are adapted for the relevant wavelength range. These goggles absorb laser beams reliably – so reliably, in fact, that the technician cannot see the beam at all while wearing them, which makes the process of making adjustments quite inconvenient!

Summary

From the point of view of physics, non-ionising radiation is not fundamentally different from X-ray and γ radiation. For frequencies below 10^{16} Hz, electromagnetic radiation is non-ionising, but biological processes in the human body are also influenced by non-ionising radiation. Low-frequency radiation (alternating current, 50 Hz) affects humans via its electric and magnetic fields. Higher-frequency radiation (GHz range) leads to heating of the tissue. Ultraviolet light can lead to a burning of the skin and even to skin cancer because this radiation is completely absorbed at very shallow depths in the tissue. Special care has to be taken with laser beams, as they have particularly high power density.

Chapter 12
How Can We Stay Safe?

Insisting on perfect safety is for people who don't have the balls to live in the real world.

Mary Shafer 1947–

In the early days of the study of radioactivity the concept of 'radiation protection' did not even exist (and there were certainly no 'radiation-protection officers'!). Physicists like Becquerel, Curie and Hahn handled relatively large quantities of radioactive substances with their bare hands. In addition to this, any radioactive material given off as gas or airborne dust was frequently inhaled. Even today the logbooks of Marie and Pierre Curie are contaminated by radium (half-life 1600 years), and they can only be loaned out from the Bibliothèque Nationale in Paris with special conditions.

Health problems caused by exposure to ionising radiation were first observed very early in the study of radioactivity. One of the first coordinated efforts to organise radiation protection was the 1915 resolution of the British Roentgen Society to protect people from overexposure to X rays. At that time, no reliable, reproducible measurement of the radiation dose was available, but they were able to observe the so-called erythema dose: this is the amount of radiation which, applied to the skin, makes it turn red temporarily. They proposed a dose limit of 1 % of the erythema dose per day, which corresponds to about 700 mSv per year. Over the course of time, the limits for exposed workers have been continually reduced, with the current level being 20 mSv per year (in the EU and in much of the rest of the world). Some practical safety measures are needed, to ensure that overexposures for workers and the general public are prevented. It is the aim of these safety measures to avoid unnecessary ingestion and inhalation of radioactive material ('incorporations'), as well as radiation exposures and contaminations.

In their simplest form, the best safety measures for working with radioactive sources are: keep as much distance between yourself and the source as possible; use shielding where applicable; work with sources whose activities are as low as is practical; restrict the time working with or near the sources; and avoid any unnecessary contaminations or incorporations.

© Springer International Publishing Switzerland 2016
C. Grupen and M. Rodgers, *Radioactivity and Radiation*,
DOI 10.1007/978-3-319-42330-2_12

To a certain extent, of course, there are radiation exposures which are unavoidable, but the aim is to reduce these unavoidable radiation exposures, contaminations and incorporations to a level As Low As Reasonably Achievable. This is the so-called ALARA principle. There are, however, some national radiation-protection regulations which have the more stringent requirement that the radiation exposure be kept "as low as possible". It is the main task of the radiation-protection officer (see Appendix B) to ensure that these requirements are fulfilled.

The topics discussed in this chapter are the practical aspects of radiation protection:

- the storage of radioactive waste, and the potential for transmutation;
- the arrangements for packaging and transporting radioactive material, the storage and security arrangements for radioactive substances in the course of normal use, and the handling of unsealed radioactive sources;
- the arrangements for mitigating the consequences of a serious accident, and for fire fighting; and
- the protection of air, water and soil from contamination.

"We store β^+-emitters next to β^--emitters so that they annihilate one another."

© by Claus Grupen

12.1 Waste Storage

If possible, radioactive waste should be reused. This might sound like an odd idea, but, for example, steel, which cannot be further decontaminated, can be reused to fabricate containers for waste of very high activity. If it is impossible to reuse some waste, it must be disposed of once and for all. Appropriate disposal of radioactive waste is a very important matter. Uncontrolled dumps are not acceptable. Radioactive waste originates mainly from the process of recycling of used fuel rods. The problem of short-lived radioactive material is essentially solved by waiting until these substances have decayed. The decay of the waste is likely to produce a substantial amount of heat: this heat emission must be taken into account and, if necessary, cooling must be provided.

Storage of radioactive waste in underground cavities is possible. There are, however, substantial requirements for such cavities, namely: geological stability; impermeability for liquids and gases; and sufficient heat-conducting properties. Still, from the case of the natural Oklo reactors (see Section 8.4), we know that it is possible to maintain containment of nuclear waste over very long periods of time.

The problem of existing waste is vast: for example, the amount of radioactive waste from the former Soviet Union is about 610 million cubic metres with a total activity of 10^{20} Bq. The fact that this is an order of magnitude higher than the amount released in the Chernobyl disaster demonstrates the scale of the potential danger if this stockpile is not dealt with adequately. It is worth noting that the amounts of waste produced from earlier generations of nuclear power plants, and also from weapons production, are much greater than the quantities produced by modern plants. This means that there is little disadvantage for those countries that have already accumulated nuclear waste in constructing new nuclear plants, as the additions to that stockpile will be marginal.

For safe storage of nuclear waste one also has to consider that radiation emerging from the waste could change the properties of the material of containers. Take as an example storage in steel containers: problems may occur over a period of several decades or centuries. The ionising radiation from the radioactive waste will interact with the steel container, and this irradiation will modify the container's properties over the course of time. Due to the continuous bombardment of the steel's structure with ionising radiation, some atoms will be displaced from their original position (so-called lattice defects). These defects can have several effects: they can make the steel brittle, cause it to corrode, change its colour, and increase its hardness. One of the benefits of steel is that, unlike other materials exposed to neutron radiation, it does not tend to become activated (i.e. to become radioactive itself). These effects have to be considered when containers and cavities for final disposal of nuclear waste are being designed and planned. It is generally believed that the common method of storage in steel containers, due to the expected changes in the material properties, guarantees safe storage only over a period of about 100 years.

There is, however, a solution to this, which is to use a different material: specifically, a material can be used which has a disordered structure from the start, and is still not brittle or unstable. In this kind of material, further atomic displacements due to irradiation will not reduce the durability, and so it would not be expected that the containers would start to leak after some time, as steel ones do. One such example is erbium zirconate. These radiation-resistant compounds could lead the way to safe long-term storage of highly radioactive nuclear waste.

12.2   Waste Transmutation

In principle a more elegant way for the handling of radioactive waste can be envisaged, namely, the transformation of undesirable isotopes into acceptable ones, by neutron or proton irradiation. This process is called transmutation. It is the long-lived radioisotopes that are the main problem in nuclear waste storage, due to the very long storage times required before they are stable: for example, neptunium-237 has a half-life of 2 million years. The drive is to transform these long-lived isotopes

into short-lived or even stable ones. To demonstrate how this could work, let us consider the dangerous isotope strontium-90 (half-life 28 years). It could be transformed
by proton irradiation into yttrium-90:

$$p + {}^{90}_{38}\text{Sr} \rightarrow n + {}^{90}_{39}\text{Y}.$$

Yttrium-90 is a short-lived β-emitter (half-life 64 hours), which in turn would decay
within a few weeks into stable zirconium-90.

However, ^{90}Sr can also be safely stored for 100 years, after which its activity
has been reduced considerably by decay. The real problems come from the very
long-lived isotopes. In recent years, significant progress has been made in the field
of transmutation of long-lived radioisotopes, although not (directly) by proton irradiation, as explained below.

During the running of a nuclear power plant, some elements heavier than uranium
(transuranic elements) are produced, when multiple neutrons attach to the uranium
and there are some β-decays. The main transuranic elements produced are plutonium,
neptunium, americium, and curium. Even though the transuranic elements represent
only about one per cent of the total radioactive waste, the problem arises because of
their extremely long half-lives (for example, plutonium-242 has a half-life of nearly
400 000 years).

For these elements, proton irradiation is not productive: instead, they would need
to be bombarded with fast neutrons. This would either fission the nuclei or transform
them into short-lived isotopes. Typical transmutation products are isotopes of ruthe-
nium and zirconium, which are either stable or relatively short-lived, with half-lives
up to a year. The safe storage of these isotopes only has to be guaranteed for a period
of a few decades. It is believed that salt domes could ensure this. It is hard to imagine
a man-made facility which can maintain safe storage for millions of years.

Unfortunately, the probability of these transmutation reactions (per time a neutron passes the radioactive nucleus) is rather low, so long irradiation times with large numbers of neutrons are required. Large concentrations of fast neutrons can be generated by a spallation neutron source (SNS). This is a system in which energetic protons from an accelerator collide with a heavy target (e.g. lead) and produce a large number of fragments, including many neutrons. In this way a single proton can liberate 30–50 neutrons in a collision. Clearly, this technique requires a proton accelerator. As early as 1984, this method was demonstrated to work at the European Centre for Particle Physics (CERN) in Geneva, Switzerland. In this experiment, small quantities of plutonium were converted by transmutation into short-lived fragments.

Transmutation by spallation neutrons, created by protons from an accelerator, is relatively simple, at least in principle. The process of transmutation cannot get out of control, because external neutrons have to be continuously delivered to keep it running. If the proton beam stops, the transmutation stops immediately. There are, however, some disadvantages and a number of unsolved problems. An expensive, high-power proton accelerator is required, and the sample of transuranic elements must be of high purity (perhaps 99.99 %). It is not possible with current technology to produce samples of transuranic elements which are this pure. The efficiency of transmutation is only about 20 %, so in order to bring down the amounts of these elements to an acceptable level, there would have to be multiple stages of separating and irradiating the material. In 2012, there was a breakthrough in nuclear waste transmutation technology by researchers from Belgium's Nuclear Research Centre in Mol and nuclear physicists from France. They successfully operated a research reactor initiating transmutation reactions with a beam of neutrons from an SNS.

As an interesting side effect, it could be profitable to use the heat generated in the process of transmutation as an energy source. Supplying the world's energy by burning fossil materials like coal, gas, and oil is not a good idea in the long run, because of the expected adverse effects on the climate. The present availability of alternative energy resources (such as solar power, wind and geothermal power) also needs to be improved.

© by Claus Grupen

At the time of writing, 437 nuclear power plants worldwide are in operation, and they will continue for some time yet. They, and their reprocessing facilities, are producing about 10000 tons of nuclear waste per year. This means that the issue of dealing with nuclear waste is unavoidable, and if there were a technologically tested method to transform nuclear waste into material which can be safely stored, it would certainly be a convincing addition to discussions about the necessity for and practicability of nuclear energy with inherently safe nuclear reactors. A large-scale implementation of this new transmutation technique would require considerable development and implementation effort. The Multipurpose Hybrid Research Reactor for High-tech Applications (MYHRRA) in Belgium, whose construction is expected to start in 2017, is a promising candidate for transmutation studies. It could demonstrate the feasability and cost-effectiveness of nuclear transformation.

12.3 Packaging and Transport

It is mandatory to ensure safe transport of radioactive material in an appropriate container. Sources must be sealed for transport, so if the source itself is not sealed, then the container it is transported in must be. The transport of radioactive material has to be accompanied by documentation in which it is stated which radioisotope is being transported and what its activity is. The material is classified by a transport class number, which depends on the maximum dose rate over the outside surface of the container. Containers must be properly labelled (see Figure 12.1).

The transport of CASTOR containers (see Figure 12.2) from nuclear power plants to recycling facilities has gained a large amount of publicity. CASTOR is an abbrevi-

Fig. 12.1 Example of a label for a transport container, showing its radioactive inventory: it contains transport material of safety category III (in red letters), specifically the isotope ^{137}Cs, an emitter of both βs and γs, with an activity of 10 MBq. The transport index 2.0 indicates a maximum dose rate at the surface of 2.0 mSv/h, and the number '7' refers to a wider system, in which radioactive material is class 7

Fig. 12.2 CASTOR-like transport and intermediate storage container for spent fuel elements. (Image credit: www.bam.de)

ation for CAsk for Storage and Transport Of Radioactive material. CASTOR containers for fuel elements are typically 6 m long and they have a wall thickness of 45 cm. The exterior of the container is ribbed, which increases the surface area, and allows the container to emit heat very effectively. CASTOR containers are thoroughly tested for all kinds of accidents possible during road or rail transportation. For example, the CASTOR container must withstand a fall from large height (9 m) onto a concrete floor. It must resist a fire at 800 °C, and must remain intact after a collision with a train, a truck or a concrete wall, or even the impact of a projectile.

Normally, CASTOR containers contain radioactive material with an activity between 10^{17} Bq and 10^{18} Bq, corresponding to 16 fuel rods. The dose rate at the surface of a CASTOR container is limited to 10 mSv/h, and the limit is 2 mSv/h at the outer edge of the transport vehicle. At a distance of 2 m, the dose rate must be smaller than 100 μSv/h. In practical cases, the dose rates measured fall significantly below these legal limits.

The loading of CASTOR containers with spent fuel rods follows a specific process, as follows. Inside the core of a nuclear reactor, cooling water also acts as shield for the radiation emerging from the fuel rods. It is inevitable that in this process the water will be severely contaminated. During the CASTOR container loading procedure, which is done under water for shielding reasons, the outside of the container must be protected from contamination by the contaminated water. Therefore it is enclosed in a protective cover which is filled with pure, uncontaminated water. After the container has been loaded, it is closed while still under water. After retrieving the container from the water pool, the contaminated water still inside the container is drained through holes which are then sealed with specialised bolts. Subsequently the outer surface of the container is cleaned with uncontaminated pure water and dried. After this, a wipe test ensures that the surface contamination of the container does not

exceed the allowed limits. The average surface contamination of the containers is normally below $0.4\,Bq/cm^2$, and it is forbidden that any single point should exceed an activity of $4\,Bq/cm^2$.

Surface contaminations on specific parts of the container exceeding these limits can easily originate from contaminated water which is sitting in cavities or in the areas around the closed drainage bolts. The contamination may spread over larger areas during the transport. This has happened in reality, and it was suspected that the containers were not sufficiently well sealed. However, the containers were actually watertight and residual contaminations originated from contaminated water not being completely wiped off.

As well as the potential of surface contamination to cause a direct exposure, it is important to consider the possibility that radioactive material from the surface could be incorporated by the handling or transport staff. However, the containers are enclosed in a casing for transport so that direct access to the contamination is extremely improbable. Only when these containers are being loaded with fuel rods, or after they reach their final destination, is the casing removed from the containers, but this procedure is performed by qualified personnel who are trained to respect the relevant safety rules.

12.4 Storage and Security of Radioactive Substances

Radioactive sources must be stored in safes, or in protected rooms or containers, while they are not being used. Particular care has to be taken against loss and theft. The radioactive material should not be stored with other non-active material.

A very important point in the storage of nuclear fuel is that the vessel containing the radioactive material must be made in such a way that it is totally impossible for the fuel to reach a critical state while being stored. From nuclear physics it is known that in the fission process, about two or three neutrons are liberated per fission (see Section 8.1). This is used in a reactor to operate the plant with constant power, by having (on average) one of the liberated neutrons initiating further fission reactions, and the other one or two not. In this case the reactor is said to be in critical state.[1] This state must be prevented under all circumstances when nuclear fuel is being stored. If a certain quantity of nuclear fuel is already being stored in a protected room or a container, and then additional nuclear fuel is added to it, it may reach a critical state with the danger of causing uncontrolled fission. In the case of storage of nuclear fuel, it is advisable to have radiation monitors measuring the dose rate and, in particular, the dose rate due to neutrons.

12.5   Handling of Unsealed Radioactive Sources

It will sometimes be necessary to handle unsealed radioactive sources whose activity exceeds the limits given in the radiation-protection regulations. In these cases, it is necessary to make sure that procedures are adopted so that the incorporation of radioactive substances and contamination of individuals involved are kept as low as reasonably achievable. The individuals handling unsealed sources may have to wear protective clothing, use protective equipment, and use respiratory protection. It goes without saying that they must avoid incorporating radioactive substances or bringing them dangerously near to the body. In particular, eating, drinking, and smoking are forbidden if unsealed radioactive substances are being handled.

Unsealed sources may only be left in the working area as long as they are actually being used. Otherwise they have to be locked up in a safe in order to reduce the possibility for incorporation or contamination.

During the handling of unsealed radioactive substances, some contamination may not be completely avoidable. It must be checked whether people have been contaminated while working with these unsealed substances. This has to be done when they leave the restricted areas in which these unsealed substances have been handled. If contamination of the skin is detected then appropriate steps have to be taken immediately to prevent a danger of further dispersion or of incorporation.

Any decontamination must be performed only by qualified personnel who have the appropriate training. Items of clothing which may be dangerously contaminated must be kept in the restricted area. Safe storage of these objects has to be guaranteed: they may even have to be treated as radioactive waste. Workplaces which have been contaminated during the handling of unsealed radioactive substances can only be used for other purposes if they have been thoroughly decontaminated.

---

[1] If more than one neutron per fission initiates a new fission (even a tiny fraction more than one on average), the reactor is said to be supercritical, and the reaction will continually increase in strength.

12.6 Mitigating the Consequences of Accidents

The most serious kinds of accident are severe accidents and design-basis accidents. A severe accident is defined by the International Atomic Energy Agency as one which leads "to significant consequences to people, the environment or the facility", with examples being "lethal effects to individuals, large radioactivity release to the environment, or reactor core melt." A design-basis accident is one which is caused by a flaw in the design of the equipment associated with the radioactive material. In any restricted area, it is compulsory to maintain, at all times, both the necessary number of staff and the required equipment for localising and eliminating the dangers connected with severe and design-basis accidents. The authority responsible for public safety must be provided with plans to eliminate or to mitigate the consequences of accidents and emergencies.

If an emergency situation does arise, the population which might be affected must be informed. This communication should be coordinated with the authority responsible for public safety. The information for the public must be repeated and updated if this is required by the emergency situation. The threshold level for a radiological emergency situation varies from country to country. Exposures of more than 50 mSv for the public would certainly create an emergency situation.

12.7 Arrangements for Fire Fighting

Appropriate fire-fighting arrangements must be planned: in particular, it is important to specify those areas where the fire brigade can work without special protection from the hazards of radioactive material (Hazard Class 1); those areas in which the

Fig. 12.3 Fireman with
protective clothing and
respiratory equipment.
(Fire Brigade Siegen, photo
taken by Claus Grupen)

fire brigade can only work if they have both safety equipment (see Figure 12.3) and
a system of health monitoring (Hazard Class 2); and, finally, areas where the fire
brigades can only work with safety equipment, health monitoring, and the specific
supervision of a radiation protection health physicist (Hazard Class 3).

For Hazard Class 3 areas, a radiation expert also has to decide which preventive
measures are to be taken to reduce the exposures to within acceptable limits. These
acceptable limits are defined by the fire brigades themselves. As an example, firemen
in Germany are only allowed to receive once in their lifetimes a radiation dose of
250 mSv in a radiological emergency situation. The total dose received must stay
within a limit of 400 mSv over the whole working life of a fireman.

12.8 Protection of Air, Water, and Soil

Nuclear installations, nuclear hospitals, and nuclear power plants will very likely discharge certain amounts of radioactive substances into air, water, or soil. The national regulations will give clearance levels for possible contaminations of the environment, i.e. they will define levels of contamination which are sufficiently low that they can be permitted. In addition to clearance levels, there are limits for the inhalation of gaseous radioactive substances and for ingestions of radioactive material which might find its way into food or drink, for example.

There are defined (and low!) limits on the concentration of artificial radioactive materials in air, water, and soil. If a maximum value is exceeded, then the operators of the relevant plant must ensure that emissions are reduced subsequently. In particular, the body doses for the public must be smaller than the specified limits on average over a period of three successive months. These limits are provided for all radioisotopes, separately for possible discharges into air, water, and soil. These limits are guided by the principle that no individual in the population may receive a dose of more than 0.3 mSv per year from these discharges, in total.

Summary

Practical aspects of radiation protection require the existence of suitable safety techniques. Laboratories must have strict procedures in place, and the disposal of radioactive waste must be performed with care, and with all relevant standards being followed. Some very simple rules can be identified: keep a good distance from the radioactive material, use appropriate shielding, restrict the time working in the presence of the material, limit the activity of radioactive substances used, and, if possible, avoid contamination, ingestion and inhalation, or at least reduce them to a minimal level.

Chapter 13
What Have We Learned?

The only real security that a man can have in this world is a reserve of knowledge, experience and ability.

Henry Ford 1863–1947

In this book, we have discussed exposures to radioactivity, and tried to describe where they come from and to put them in context. Everything is radioactive to some extent, including our own bodies, and so radiation exposures are part of normal life. The human race evolved alongside this background radiation, which has sometimes been at a level several times higher than it is at the moment, and it has very rarely caused us problems, which is why we never evolved a sense for this radiation. In the past century, the human race has learned how to put radioactivity and radiation to good use for nuclear power, for nuclear medicine and in industry.

In order to talk about exposures, we need a sensible scale, and so we use the number of millisieverts that have been received. We have discussed natural and man-made sources of radiation: in thinking about these, it is important to remember that, as far as the effects are concerned, there is no difference between a millisievert from natural cosmic rays, a millisievert from a medical procedure, and a millisievert received by a worker making radioisotope batteries.

In dealing with radiation, it is always a matter of degree: the question is whether sensible limits are being respected. For members of the public, limits are so low that exposures are not noticeable. Radiation workers have higher limits, but their doses need to be painstakingly monitored and minimised. There have been many times, in the civilian and military sectors, that basic safety rules have been disregarded, and excessive doses have resulted for some individuals, but fortunately, these incidents have mostly been on a small scale. Even the rarest, most catastrophic accidents (Chernobyl and Fukushima) have killed relatively modest numbers of people if compared with other major disasters, such as famines and earthquakes. In the case of Fukushima, the tsunami which led to the nuclear emergency caused hundreds of times more destruction and death than the nuclear disaster itself. It is also worth

© Springer International Publishing Switzerland 2016
C. Grupen and M. Rodgers, *Radioactivity and Radiation*,
DOI 10.1007/978-3-319-42330-2_13

remembering that an old, poorly-maintained nuclear power station will still emit less radioactivity than a modern, relatively clean coal-fired one.

One way of thinking of radioactivity is that it is like salt, but less bad. Getting absolutely no salt in our diet is impractical, and probably not desirable. A normal, healthy diet contains (say) 2 grams of salt per day. A normal background radiation dose is about 0.007 mSv per day. A person receiving a radiation exposure a thousand times this level, 7 mSv per day, would be likely to experience some negative long-term effects after a few years. But a person eating a thousand times the normal amount of salt, 2 kg per day, would die much sooner!

Our lives abound with potentially dangerous circumstances too numerous to mention, from the obvious (like mountaineering or drinking large amounts of alcohol) to the mundane (crossing the road or being overweight). For comparison, even if we take the standard estimate of risk seriously (it is likely to be a large overestimate), the typical annual radiation exposure of the public from nuclear power plants has less risk associated with it than eating one portion of red meat,[1] or crossing the road a few times (as regards being injured), in that year.

We hope you have found this book informative and useful, and if you would like to learn more, please have a look at the books listed under Further Reading.

With a little knowledge, there is no need to be this scared.

---

[1]The American Institute for Cancer Research concluded in 2007 that "red or processed meats are convincing or probable sources of some cancers". Clearly, the risk from one portion is extremely low, which is the whole idea of this comparison.

Appendix A
How Can We Detect Radiation?

> *If our tongues were as sensitive as these radiation detectors, we could easily taste one drop of vermouth in five carloads of gin.*
> Dixy Lee Ray 1914–1994

It is impossible to smell, see, taste, hear or feel ionising radiation. Humans have no senses for α, β and γ rays, so detectors are needed, to play the part of this absent capability – to sense ionising radiation for us. It is necessary to measure radiation exposures so that they can be monitored, controlled and limited. Humans also have to be protected against unexpected exposures. One important duty is to supervise radiation-exposed workers, and to measure external radiation exposures, contaminations and incorporations, particularly in working areas. Another is to protect the environment against unnecessary exposures. This includes the determination of radiation exposures of the general public, the monitoring of the disposal of radioactive waste into the environment, and the examination of the distribution of radioactive material in the biosphere (the atmosphere, soil, water, and eventually food).

National radiation-protection authorities have realised that radiation exposures from natural sources also have to be considered. In certain situations, natural radiation can increase the radiation level for particular individuals in the population quite considerably. Therefore, these natural sources cannot be neglected in the framework of radiation protection.

The detectors used in the field of radiation protection have to be reliable and robust, and their measurements have to be reproducible. It is important to note that for each of the various types of radiation, an appropriate detector needs to be used. This chapter introduces the various types of radiation detector normally used in industry, academia and elsewhere.

A.1 Ionisation Chambers

An ionisation chamber is a detector with a gas-filled region, which measures the amount of ionisation produced by a charged particle passing through the gas. Neutral

© Springer International Publishing Switzerland 2016
C. Grupen and M. Rodgers, *Radioactivity and Radiation*,
DOI 10.1007/978-3-319-42330-2

Fig. A.1 Sketch of an
ionisation chamber

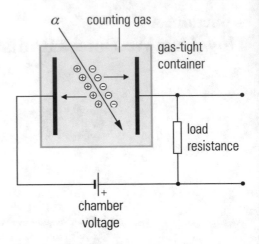

particles can also be detected by this device via secondary charged particles (those that result from the interaction of the initial neutral ones with electrons or nuclei). The charged particles move through the gas, and strike gas molecules, liberating electrons and leaving ions: i.e. they produce more charged particles. The number of charged particles can be measured using an electric field, which attracts the oppositely charged particles in opposite directions onto conducting plates (the electrodes), from which the signals caused by the particles can be recorded.

In the simplest case, an ionisation chamber consists of a pair of parallel electrodes mounted in a gas-tight container that is filled with a gas mixture (see Figure A.1). A charged particle produces electron-ion pairs along its track, and the electrons drift through the gas towards the positive electrode, while the ions drift towards the negative electrode. The number of charge carriers created depends on the particle type and energy, and on the gas used.

The parallel electrodes of the ionisation chamber are initially charged to a certain voltage, producing a homogeneous electric field, i.e. one which is the same strength and points in the same direction at all positions within the field. The drifting charge carriers induce an electric charge on the electrodes, which leads to certain change of the voltage. This voltage change has contributions from the much lighter, fast-moving electrons and from the much heavier, slowly drifting ions. There is a significant disadvantage, which is that the output signal depends on where the particles enter the ionisation chamber. This can be solved by introducing a Frisch grid (a third electrode, constructed to reduce the dependence), or by only permitting the particles to enter the chamber from a well-defined region. With either remedy in place, the amount of energy deposited by the particle can be deduced from the signal: and if the incoming particle is totally absorbed in the gas, the detector measures its total energy.

Ionisation chambers are mainly used to detect particles with high charges, because in this case the energies deposited tend to be larger than those of singly charged

"The detector is that small because it has to detect tiny particles"

particles. α particles have a charge of two units, and if coming from a radioactive source, they will typically deposit 5 MeV in a 4 cm layer of argon, which is easily detected. A singly charged particle, such as a β particle, passing through the same 4 cm layer of argon will deposit only about 11 keV, which provides about 400 electron-ion pairs. To detect such a small signal is a very difficult task!

Figure A.2 shows a parallel-plate ionisation detector. The charged plates, which are on the top and bottom of the large square area, are sufficiently thin that they are transparent. The controls (for setup, reset and test mode) are next to the built-in display. The device can also be connected to a computer, so that readings can be logged.

As well as parallel-plate ionisation counters, cylindrical ionisation counters are also in use. These have the positive electrode (anode) as a thin wire along the centre of the cylinder, and the shell of the cylinder as the negative electrode (cathode). Because of this cylindrical arrangement, the electric field in this case is no longer homogeneous, but is stronger closer to the anode wire and weaker towards the edges of the cylinder.

The important point with ionisation chambers is that no multiplication of charge carriers occurs in the gas. Only the electrons and ions originally produced by the incoming particle are collected.

Fig. A.2 A parallel-plate
ionisation chamber. (Model:
DIAMENTOR SET CX,
from PTW, Freiburg,
Germany)

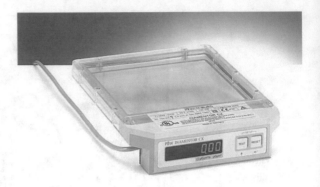

A.2   Proportional Counters

Proportional counters are in principle similar to ionisation chambers, but they use a
higher voltage between the electrodes, giving a stronger signal due to an amplification
effect. When a particle enters the container, it will ionise atoms and produce pairs
of positive ions and (negative) electrons. These particles are accelerated strongly
in the powerful electric field – sufficiently strongly that they themselves produce
further electron-ion pairs when they strike other gas molecules. This process can
continue with future generations of charged particles. These cascades of particles
can eventually be measured when they strike the electrodes. Crucially, as long as
ionisation remains the only significant process, the signal measured is proportional
to the original amount of ionisation, but multiplied by, typically, 10^4 or 10^5.

Proportional counters have a natural upper limit to their operation: at yet higher
voltages, the accelerated electrons will not only ionise gas molecules, but will sometimes excite them (put molecular electrons into higher energy states, without ejecting
them from the molecule). De-excitation usually involves the emission of photons,
which can interact with the electrodes (by the photoelectric effect), producing further
charged particles and amplifying the cascade. The addition of this effect means that
the counters are no longer proportional for very high voltages, and so they are not
used with voltages over that level (which depends on the device, but is usually about
1000 V).

Since the output signal in a proportional counter is proportional to the energy
loss of the particle (or to the total energy of the particle, if the particle stops in the
counter), these radiation detectors can be used for spectroscopy, i.e. to determine the
distribution of energies for a group of many particles (e.g. of X rays or α particles).

A multi-wire proportional chamber (MWPC) is a more advanced device, operating
much like a proportional counter, but with multiple charged wires across a planar
chamber. In additon to signals proportional to the energy, MWPCs provide accurate
track information if the wires are stretched with relatively small separations. Then

major use as regards radiation protection is as contamination monitors: measuring the extent of contaminations over large areas.

A.3 Geiger-Müller Counters

For many circumstances, a strong detector response, even to low levels of signal, is important, even if the cost is that the energy deposited by the particles cannot be measured. In a Geiger-Müller counter (also called a Geiger counter or a GM counter), a still stronger electric field is used, so that the output is no longer proportional to the energy left in the chamber by the passing particle. Like some ionisation chambers, the shape of a GM counter is a cylinder, with the wall negatively charged and a central wire positively charged. Each ionising particle passing through the chamber causes a single brief discharge, and the count of these discharges gives an indication of the radiation level. In order to turn that indication into a measurement of dose, a different calibration is required for each radiation type and energy.

Increasing the field strength in a proportional counter, beyond the limit where it ceases to be proportional, leads to copious production of photons during the formation of the particle cascade. These photons, often in the ultraviolet range, may knock electrons out of the electrodes if they strike them (the photoelectric effect). The photon may have travelled some distance (typically several millimetres) before striking the electrodes, and the electrons liberated can cause further cascades, distant from the original interaction point. Another effect with a similar consequence is the strong acceleration of the gas ions: at these energies, they can strike the walls of the cylinder (negative electrode) hard enough to cause ionisation there, also far from the initial interaction point. These effects create multiple signals for one initial interaction, and make the cascade last longer so that several cascades can run into each other: the measurements of the counter would be meaningless if this were allowed to happen.

This problem of a spreading, continuous discharge can be stopped by so-called quenching, making a Geiger-Müller counter. This is the addition of a gas with relatively large molecules (the 'quench gas') into the counting gas (which is in most cases a noble gas). The quench gas molecules have two effects. Firstly, they absorb the ultraviolet photons, typically after the photon has travelled 0.1 mm. Secondly, they neutralise the positive ions of the counting gas by giving up electrons to them. Because the quench gas molecules are larger, the positive charge is spread more thinly across them, particularly in relation to their masses. This means that the quench gas ions only accelerate relatively gently even in the strong electric field, and do not cause ionisation when they hit the cylinder walls. In combination, these effects restrict the size of the cascade, and stop the production of secondary signals. This means that each count clearly represents a single incoming particle, and the detector is ready very soon afterwards to make a further count if another particle enters. It does not, however, leave any information about the energy or initial position of the incoming particle.

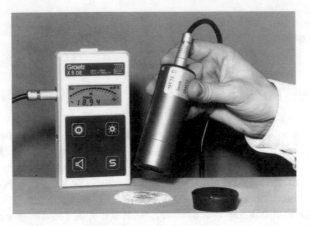

Fig. A.3 Taking a measurement of activity with a Geiger-Müller counter. (Tube 18526D and counter X5DE, from GRAETZ, Altena, Germany)

There are many gases which will act as quenchers: hydrocarbons such as methane (CH_4), ethane (C_2H_6) and isobutane (iC_4H_{10}); alcohols and related compounds, such as ethyl alcohol (C_2H_5OH) and methylal ($CH_2(OCH_3)_2$); and halides, particularly ethylbromide (C_2H_5Br).

Figure A.3 shows a measurement being taken with a Geiger-Müller counter. The tube is of a convenient size to be hand-held, and the display (after appropriate calibration) gives the dose rate in mSv/h or μSv/h. In this case the measurement is of the activity of a gas mantle, which is a component of some old-fashioned lighting systems, and contains mildly radioactive thorium. Thorium is chosen for this task because it emits visible light efficiently when heated: its radioactivity is not used for lighting. The mantle is lying on the table, and looks like a small piece of netting.

"Counter"

© by Claus Grupen

A.4 Solid-State Detectors

Solid-state detectors are essentially ionisation chambers with solids as a counting medium. Because of their high density compared to gaseous detectors, they can absorb particles of correspondingly higher energy. Incoming charged particles or photons promote electrons into higher states in the crystal (excitation). An electric field applied across the crystal allows the charge carriers produced by this excitation to be collected.

The operation of solid-state detectors can be understood from the so-called band model of solids. This relies on the solid being a crystal, i.e. being an ordered structure (think of a three-dimensional grid) containing many trillions of atoms. The crystal has certain energy levels (bands) which can each take a certain number of electrons. Some low energy bands are fully filled with electrons while the high energy bands are empty, at least at room temperature. The highest-energy of the fully-filled bands is called the "valence band", and the next band up from that, which is the lowest-energy partially filled or empty band, is called the "conduction band". The energy difference between these bands is called the band gap (or sometimes the "forbidden band").

When the conduction band is partially filled, electrons can move easily under the influence of an electric field, hence, this solid is a conductor. Such a material cannot be used as an ionisation counter. The solids which have effectively empty conduction bands are divided conventionally into insulators (in which it is very difficult to promote electrons into the conduction band at room temperature) and semiconductors (in which this promotion is relatively easy). The electric charge in these materials is carried by electrons, which have been excited to the conduction band from the valence band. The corresponding vacancies in the valence band are called holes and are able to drift in the electric field as well.[1]

The transfer between the valence and the conduction bands in a semiconductor can be made yet easier by "doping" the crystal: adding certain well-chosen impurities.

If a charged particle traverses a doped crystal, it will produce electron-hole pairs along its track. The electrons released can also collide with other parts of the material and produce further electron-hole pairs, or they can cause vibrations inside the lattice. A channel is produced along the particle track with charge-carrier concentrations of something like 10^{16} per cm^3. The solid-state detector will then be effective if it is possible to collect the mobile electrons using an external field before they can recombine with the holes. If this is successful, the charge signal measured is proportional to the energy loss of the particle or, if the particle deposits its total energy in the sensitive volume of the detector, it is proportional to the total particle energy.

Solid-state counters can be used to find specific energies of α, β and γ rays, and to count their numbers accurately. High-purity germanium detectors, like the one in Figure A.4, provide precise energy measurements for photons with energies in the MeV range. The precision can be as good as 0.1 %, allowing the specific isotope to

---

[1] Or rather, everything else in that band drifts in such a way that it looks like the hole is drifting.

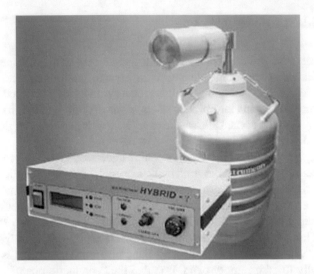

Fig. A.4 A germanium detector. The germanium crystal is the flat circular plate, and the flask underneath the detector (called a dewar) contains liquid nitrogen for cooling. (Image credit: NATS, Conneticut, USA, nats-usa.com/high-purity-germanium-planar-spectroscopy-system)

be identified by its radiation, as in Figure A.5. If the detectors are segmented into strips or pixels, it is possible also to achieve outstanding spatial resolutions, down to $10\,\mu$m. This, however, requires a large number of readout channels equipped with sensitive amplifiers. Silicon pixel detectors suffer radiation damage, and so ageing effects must be taken into account if they are used in a harsh radiation environment.

It is also possible to use diamond for this type of detector. Diamond detectors have a relatively large band gap, and so only high-energy radiation can cause an electron to be promoted and read out: this means that visible light causes almost no signal in them. Diamond is also much more resistant to radiation damage than other materials, so diamond detectors can be used in very harsh radiation environments, such as in some types of hadron therapy, or in the innermost parts of particle accelerators.

If yet better energy resolutions are required, and the intensity of radiation is not particularly high, cryogenic detectors can be used. These use the excitation of very low energy quantum states (phonons and Cooper pairs) in a material to measure small energy depositions. Because the differences between the energy levels are tiny, it would be very easy for thermal noise to drown out the signal, and therefore the detectors are made to operate at very low temperatures, typically a few thousandths of a degree above absolute zero ($-273.15\,°$C).

A.5   Scintillators

Scintillators are materials that emit light when they are struck by high energy particles (including photons). They can be inorganic crystals or organic materials (liquids, solid plastics).

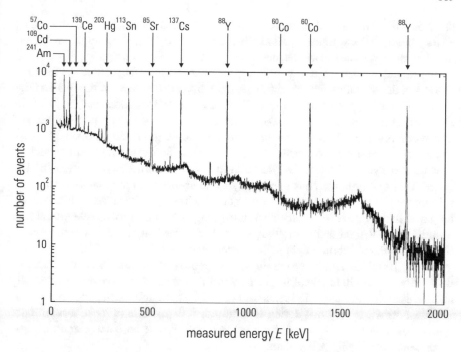

Fig. A.5 Photopeak identification from a mixture of radioisotopes using a high-purity germanium detector. In this scenario, each isotope in the mixture produces many γ photons of a specific, characteristic energy. Each photon's energy is measured very precisely, so all the photons from any one isotope appear in the same position in the graph, giving a very sharp line. Cobalt-60 and yttrium-88 have two possible transitions each, and so each has two sets of γ photons

    The scintillation mechanism in crystals is related to the band structure (described in the previous section). As in other non-conducting crystals, the electrons normally sit in the valence band. As an incident particle moves through the crystal, it gives up some energy to a number of electrons by colliding with them. When it gives a relatively large amount of energy to a single electron, it will cause ionisation, and the liberated electron will then give up energy to further electrons in the crystal. When an incident particle or a liberated electron gives a modest amount of energy to an electron in the crystal (which is the more common case), the crystal electron is promoted to a higher band (conduction band). The electron is then able to move freely through the crystal, and it may take some time before it falls back into the original state (decays), emitting light in the process. All crystals will emit this light to some extent, but those which do so most efficiently, preferably giving out visible light, are the obvious choice for scintillation detectors.

    As well as the efficiency of visible light emission, the scintillator's decay time (the time delay between an electron being promoted and the light being given out) is also important. For each application, a scintillator with an appropriate balance between these parameters is chosen. For example, thallium-doped sodium iodide crystals produce a large number of photons for the energy input (about one photon for each 25 eV), but their decay times are relatively long (about 0.2 μs). An alternative, barium fluoride, decays in less than 1 ns, but has a less efficient conversion.

In organic substances, the scintillation mechanism is different because the excitation and de-excitation occurs within a single discrete molecule, rather than a large crystal. Certain types of molecule will release a small fraction (say, 3%) of the absorbed energy as visible or near-UV photons. This process is especially marked in a set of organic substances called 'aromatics' (nothing to do with smell!), such as polystyrene, polyvinyltoluene and naphthalene. Liquids which scintillate include toluene and xylene. The primary scintillation light is normally emitted in the UV range. This UV light then has a tendency to be absorbed in the material itself: the scintillator is not transparent for its own scintillation light. Because of this, a wavelength shifter is mixed in with the scintillating material. The wavelength shifter absorbs the UV light and re-emits it at a longer wavelength (e.g. as green visible light). It is then the shifted light which is detected.

About 100 eV is required to produce one photon in an organic scintillator. The decay time of the light signal in plastic scintillators is in most cases substantially shorter than that in inorganic substances (e.g. 30 ns in naphthalene and around 2 ns in polystyrene). If very fast timing is required, it is possible to make a scintillator with an ultrashort decay time (0.5 ns) using a polystyrene base and certain organic compounds containing heavy elements.

A.6 Neutron Counters

Neutrons, like photons, can only be detected indirectly. At different neutron energies, different interactions are used to produce charged particles: these charged particles are then detected via their ionisation or scintillation in standard radiation detectors. The energy ranges covered by the particular detectors fall into three groups.

(a) Low energies (less than 20 MeV)

$$n + {}^6\text{Li} \rightarrow \alpha + {}^3\text{H}\,,$$
$$n + {}^{10}\text{B} \rightarrow \alpha + {}^7\text{Li}\,,$$
$$n + {}^3\text{He} \rightarrow p + {}^3\text{H}\,,$$
$$n + p \rightarrow n + p\,.$$

The conversion can be performed in a proportional counter using BF_3 or ^{3}He as the counting gas (or a mixture with an addition of hydrocarbons, like CH_4). Otherwise, a scintillator consisting of a solid crystal that will also convert the neutrons, such as thallium-doped lithium iodide, can be used.

(b) Intermediate energies (20 MeV–1 GeV)

At these energies, the most efficient reaction to use is the fourth one above, in which a fast neutron collides with a proton (hydrogen nucleus), giving it some energy, but there is no change to the type of any nuclei. The fast proton will then travel through the detector causing ionisation. The best detectors contain large numbers of hydrogen atoms in their sensitive volume. Hydrocarbons, such as methane (CH_4), are often used.

(c) High energies (over 1 GeV)

Neutrons of high energy initiate cascades when they strike nuclei, and so are easy to identify in hadron calorimeters, which are blocks of suitably dense material, such as iron, with scintillators to measure the signal.

Neutrons are detected with relatively high efficiency at very low energies. Therefore, for detection purposes, it is often useful to slow down neutrons by passing them through a moderator. Moderators are materials containing large numbers of hydrogen atoms: collisions of neutrons with protons (hydrogen nuclei) are very efficient at transferring energy away from the neutron. For the purposes of radiation protection it is also important to measure the neutron energy, because the biological effectiveness (and therefore the radiation weighting factor) depends on it.

A.7 Personal Dosimeters

In the field of personal dosimetry, it is important to distinguish between directly and indirectly readable dosimeters. The only common type of directly readable dosimeter is based on an ionisation chamber, and is discussed first in this section. After this, three types of indirectly readable dosimeters are discussed.

Fig. A.6 Pen-type pocket
dosimeter

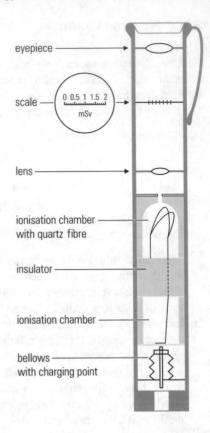

Ionisation chambers can be made into dosimeters which are small and convenient: the size and shape of a pen. A diagram of this type of dosimeter, also called a self-indicating pocket dosimeter, is given in Figure A.6. The dose received can be read directly and easily on these dosimeters. They function by showing the changes in charge on two plates with opposite charges. The interactions of incoming particles in the chamber's gas produce positive ions and (negative) free electrons, in numbers proportional to the dose being received. The positive ions are attracted towards the negative plate and move towards it until they strike it. Similarly, the electrons move towards and hit the positive plate. This causes the charge on the plates to reduce – and to do so by an amount proportional to the received dose.

Between the charged plates, there is a charged quartz fibre: the charge on this stays constant during the operation of the dosimeter. The fibre is attracted and repelled by the two plates, and so is bent whenever the plates are charged, and the extent of the bending is related to the remaining charge on the plates. With the use of some built-in lenses, the position of the fibre is shown on a scale calibrated in milli-Sieverts (or sometimes micro-Sieverts), giving the user a very direct way of observing the dose. If using these dosimeters, it is important to recognise some limitations: for example, they underreport X-ray exposure, as some X rays are absorbed in the walls of the dosimeter, and so are never detected. Furthermore, they can give somewhat variable results for γ-photon energies below 300 keV, as the probability of the photons causing ionisations in the chamber material (the photoelectric effect) is very sensitive to small variances in that energy.

The most popular and well-known indirectly readable dosimeter is the film-badge dosimeter (see Figure A.7). These dosimeters use the blackening of a photographic film (X-ray film) as a measure for the received dose. The badges have metal absorbers (often copper and lead) of different thicknesses in front of the film, so that they can give information about the different intensities of the different radiation types. Official dosimeters are mostly film badges because they give detailed information, and can be retained for permanent records. There are, however, several disadvantages of film badges, namely the limited durability of films, the sensitivity to humidity and high temperature, the limited measurement accuracy and the inconvenient method required to read the information (i.e. developing the film).

The film dosimeters described above can only determine the dose at skin-level. The next generation of film-badge dosimeters, such as 'sliding-shadow' dosimeters, is optimised for the measurement of the penetration of radiation into the skin. These dosimeters also allow a determination of the photon energy and angle of incidence. Furthermore, they can discriminate between β and γ rays.

There are two types of dosimeter which rely on delayed emission of light. These use materials which can be irradiated, and will give out light in response, but only later, when stimulated. For the first set of dosimeters, the relevant stimulus is heat (thermoluminescence), and for the second, it is exposure to UV (fluorescence). In both cases, the amount of light given off is almost unrelated to the amount of stimulus, but directly related to the amount of radiation received earlier. These dosimeters have distinct advantages: they have good sensitivity, even if very small, so they can be made into a finger-ring (see Figure A.8). Their main disadvantage is that the readout system is laborious.

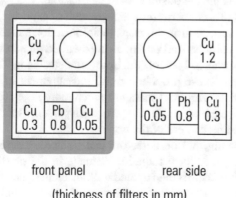

front panel   rear side

(thickness of filters in mm)

Fig. A.7 Film-badge dosimeter showing positions of different absorbers

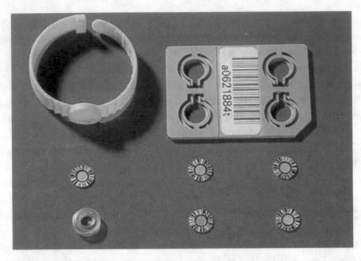

Fig. A.8 Adjustable finger-ring, with small thermoluminescence detectors which can be mounted onto it. (HARSHAW TLD, Thermo Eberline Trading GmbH)

The most common type of fluorescence dosimeter is constructed from a glass containing phosphates and a small amount of silver (and is called a phosphate-glass dosimeter). Thermoluminescence dosimeters (TLDs) use a certain set of inorganic compounds, notably lithium fluoride. There is also a new type of TLD, which uses the rare-earth metal thulium. These TLDs have particularly good performance for low levels of radiation, and have the additional benefit that they can discriminate between low-energy X-rays and γ rays.

For TLDs, the readout process (with heat) erases the previous information. This is a disadvantage in that the information is lost if there is an error in recording, and the device itself holds no permanent record (unlike a film badge or a phosphate-glass device), but it does enable the dosimeters to be reusable.

A further use of TLDs concerns the measurement of exposures due to neutrons. It can be harder to measure neutron doses, as their ionisation patterns are not as simple as those of α and β particles and photons. In so-called albedo neutron dosimeters, neutrons which are scattered back off the body are detected in TLDs (which usually use lithium or boron). These provide a purer measurement of neutron exposures, as other types of radiation are backscattered much more rarely. Unfortunately, these dosimeters have to be calibrated separately for each person, as their results are dependent on the water and fat content of the body.

For radon monitoring, plastic detectors can be used. These use the track-etch technique: α rays produce local radiation damage in the plastic material, usually cellulose nitrate. If the plastic is then treated with an alkali (etched), the tracks become clearly visible. This type of detector is quite specific to α particles, as they cause by far the most concentrated radiation damage in the material. However, the track-etch technique can be used to detect neutrons in a similar process, if a polycarbonate film is used.

Table A.1 Typical applications of different measurement techniques for personal dosimetry

Dosimeter	Method of operation	Radiation type	Measurement range	Advantages and disadvantages
Film badge	Photochemical blackening	γ, β	0.1 mSv–5 Sv	Permanent record, insensitive to low-energy γ rays
Pen-type pocket dosimeter	Ionisation chamber	γ	0.03–2 mSv, typically	Very sensitive to γ rays, continuously readable, insensitive to α and β rays, no permanent record
TLD dosimeter	Thermoluminescence measurement	γ, (β)	0.1 mSv–10 Sv	Suitable for the measurement of low doses, no permanent record
Phosphate-glass dosimeter	Photoluminescence measurement	γ	0.1 mSv–10 Sv	Permanent record, can be read repeatedly
Albedo neutron dosimeters	Neutron reflection by the carrier	n, γ	0.1 mSv–10 Sv	Calibration depends on the human carrier
Track-etch dosimeter	Material damage in polycarbonate films	n	0.5–10.0 mSv	Permanent record, the evaluation requires knowledge of the radiation type
Radon personal dosimeter	Material damage in cellulose nitrate films	α	75–7000 kBq h/m$^{3(2)}$	Permanent record, specific for α particles

[(2)]For a routine quarterly investigation (corresponding to 500 working hours) this corresponds to an average radon concentration in the workplace of 150–14 000 Bq/m^3. The lower limit of that range can be reached in poorly-ventilated domestic settings

"And these little badges are supposed to protect us against radiation?"

A summary of the various measurement techniques for personal dosimetry is provided in Table A.1. The reader will notice that there is only one personal detector for α rays in Table A.1, namely, the radon personal dosimeter. Due to the short range of α particles in air or clothing, external irradiation by α rays mostly represents almost no radiation risk. The case of radon inhalation is distinct, as the exposure to α particles is then internal, and so much more harmful. This is explained in more detail in Section 3.1.

A.8   Accident Dosimetry

Occasionally, it is necessary to determine body doses after a radiation accident, and no dosimeter information is available. In these cases, so-called accident dosimetry is required: the doses must be estimated from the evidence which is available. One possibility to work out the received body dose after the fact is given by the so-called hair-activation method. Hair contains sulphur with a concentration of about 50 mg of sulphur per gram of hair. By neutron interactions (e.g. after a reactor accident) the sulphur can be activated according to

$$n + {}^{32}\text{S} \rightarrow {}^{32}\text{P} + p \ .$$

In this way, the radioisotope phosphorus-32 is produced, which is a β-emitter with a half-life of 14.3 days. In addition to this, radioactive silicon-31 is produced by

$$n + {}^{34}\text{S} \rightarrow {}^{31}\text{Si} + \alpha \ .$$

The decays from the ^{31}Si radioisotope interfere with the phosphorus-activity measurement. However, the half-life of the ^{31}Si isotope is only 2.6 hours. Therefore, it is easy to wait until this activity is insignificant before the ^{32}P activity is measured. If there is also a surface contamination of the hair, substantial cleaning is necessary before the phosphorus activity of the hair is measured.

Because of the low expected count rates, a detector with high efficiency and low background is required. From previous experiments, the probability of making ^{32}P from any one neutron collision is well-known. This means that the radiation dose received can be inferred from the phosphorus-32 activity.

A further possibility for accident dosimetry is given by the procedure of blood activation. Human blood contains about 2 mg of sodium per millilitre. By the capture of neutrons, the stable ^{23}Na isotope transforms into radioactive ^{24}Na, according to

$$n + {}^{23}\text{Na} \rightarrow {}^{24}\text{Na} + \gamma \ .$$

Again activities of short-lived radioisotopes produced in this neutron-capture process can interfere with the ^{24}Na-activity measurement. After a suitable decay time, the remaining activity of ^{24}Na (a β-emitter with a half-life of 15 hours) can be recorded and used as starting point for an estimate of the dose received.

A.9   Overview of Radiation Detector Types

The radiation detectors presented in this chapter help to identify radioisotopes, to determine the amount of radioactivity released, to monitor possible illegal disposal of radioactive material, and to observe the effect of radiation on the environment and on humans. In typical surveillance programs for nuclear power plants, there is a requirement to determine the size of any exposures due to external and internal radiation, the release rates into water and air, which food chains might be affected, and the identities of the radioisotopes released. The corresponding measurement techniques have to be very sensitive, since the permitted limits on exposures of the general public from the release of radioactive material into the environment are usually rather low.

Dose-rate measurements for personal dosimetry predominantly employ ionisation chambers and Geiger-Müller counters. Because the outputs of these detectors vary by direction and by particle energy, and also because the types of radiation to be identified are usually unknown in advance, the measurement errors are quite substantial (20–50 %). If the activity and dose of a mixture of different γ-ray sources has to be determined, the uncertainty of the measurement can be as much as a factor of 2 (i.e. the real value could be half or double the measured one).

It is very important, as part of any radiation-protection system, to measure and monitor incorporations and contaminations. The determination of the body activity after a suspected incorporation can be performed with a whole-body counter. A measurement of the activity of biological excretions (predominantly urine and faeces) also gives evidence about possible incorporations. Since the activities for possible contaminations or incorporations are usually quite low, whole-body counters have to be properly shielded against background radiation. Finding shielding material of very low activity is non-trivial, since almost everything produced after the nuclear weapon tests in the atmosphere (mostly in the 1950 and 1960s) is slightly contaminated. To obtain material of sufficiently low activity, it is worth going to a very unusual source: steel from battleships sunk in the First World War and later recovered!

Summary

The interactions of charged and neutral particles are the basis for the development of radiation detectors. Proportional counters and Geiger-Müller counters operate using the ionisation of gases, and scintillation counters using the excitation of solid crystals or organic materials. Because films are blackened by ionising radiation, film badges provide a permanent record of the received dose. Small ionisation chambers and luminescence devices can also be used as dosimeters. If no dosimeter was present, a hair or blood sample, if tested in the right time-frame, can help estimate the scale of an exposure.

Appendix B
How is Radiation Protection Organised?

My time inside there was very short compared to the amount of time it took to take on and take off this suit and to test me for how much radioactivity I have.

William Scranton 1947–

(after visiting the Three-Mile-Island reactor)

The fundamental principles for radiation protection are laid down in recommendations by the International Commission on Radiological Protection (ICRP), and the International Commision on Radiation Units and measurements (ICRU). These guidelines have been published in a series of recommendations since 1991, with the most recent update dating from 2007.

These recommendations are based on three fundamental principles, namely: *justification* for radiation exposures, *optimisation* reducing the exposures for radiation workers and the public, and *dose limits* defining maximum acceptable doses.

The idea of justification is that the exposure to radiation should do more good than harm. In all situations where radiation sources are going to be used, one has to balance the benefit and the possible damage, and the positive must always outweigh a possible negative. Optimisation means that the application of radioactive sources must folllow the ALARA principle: the exposure to humans should be As Low As Reasonably Achievable. It is of course necessary to establish dose limits for radiation workers and the public. Dose limits, which should not be exceeded, are recommended by the ICRP.

It is important to keep in mind that the proposals by the ICRP are merely recommendations. They must be incorporated by the national authorities into national regulations before they can take effect. The various national regulatory bodies can simply adopt the ICRP recommendations, or they can impose more stringent limits, but it is not advised that they choose more relaxed ones.

The body which sets the radiation rules and guidelines, and is responsible for oversight of them, is called the competent authority. This is a different authority in each country, and is usually either a health agency or a specific radiation protection

© Springer International Publishing Switzerland 2016
C. Grupen and M. Rodgers, *Radioactivity and Radiation*,
DOI 10.1007/978-3-319-42330-2

commission. For example: in the USA, the United States Nuclear Regulatory Commission is the competent authority; in the UK it is the Health and Safety Executive; and in Germany it is a different authority in each of the 16 Länder (states).

In radiation protection, it is necessary to distinguish between three groups: radiation workers (for example staff in nuclear power plants, in research institutions or in hospitals operating radiation facilities); patients; and members of the public. The exposure for individuals working in radiation areas should not exceed 100 mSv in a consecutive five-year period, so on average a dose of 20 mSv per year should not be exceeded. A maximum effective dose of 50 mSv in any single year can be tolerated if it is compensated for by lower doses in the following years. For members of the public, the limit is 1 mSv per year.

These dose limits apply only to radiation caused by technical installations. There are not normally any attempts, using laws, to limit radiation from natural sources, i.e. cosmic rays, terrestrial radiation, inhalation of normal air and ingestion of unpolluted food and drink. This natural radioactivity amounts to about 2.5 mSv per year for the majority of people. The normal variation of this exposure for different places on

Earth was used as a guideline in the setting of the limit for the general public (at an additional dose of 1 mSv per year).

For patients, the situation is different: for some medical treatments, particularly of cancers, the exposure will certainly exceed the limits given. The radiation dose in this case is given by the rule that the cancerous cells should be effectively destroyed, but the healthy tissue should be spared as much as possible. Because a case-by-case medical judgment is required, it does not make sense to set an advised limit for patients.

The estimation of possible long-term damage from radiation is based on the *Linear No-Threshold (LNT)* hypothesis, i.e. it is assumed that possible risks of radiation-induced cancer or genetic defects are linearly proportional to the dose, and that there is no threshold for detrimental effects of radiation (see Chapter 7).

The LNT hypothesis is controversial and hotly debated, particularly the ("No-Threshold") assumption that there is no dose low enough to have zero risk. Mankind has developed under radiation from natural sources, and the level in the distant past was even higher than today. There are two reasons for this. Firstly, some naturally radioactive isotopes have decayed during the evolution of mankind, and are now producing less radiation. Secondly, the Earth's magnetic field, which protects us from most cosmic rays, varies in strength over time, so sometimes it allows a much larger cosmic ray exposure. One can also argue that mankind has adapted to ambient levels of temperature, pressure, oxygen concentration in the atmosphere, and also to the level of natural radiation. There is even a contentious hypothesis called hormesis, which holds that some low level of radiation is beneficial (similarly to the observed protective effect of low doses of alcohol). This has been observed, although not consistently, in some non-human biological systems, and is even more controversial when applied to humans (but there is an interesting human study from Taiwan, discussed in Section 7.6). Despite these points, the LNT hypothesis is useful because it provides relatively clear and conservative limits.

*"I like this radioactively enriched food.
It strengthens my immune system!"*

Another observation which is frequently underestimated is that psychological effects, believed to be unrelated to radiation exposure, result from the fear that radiation exposure could cause health damage in the future. A surprisingly high incidence of psychosomatic symptoms like headaches and depression has been found in only very slightly contaminated areas after radiation accidents. These health problems could also be due to economic hardships or evacuation after radiation accidents, and not to the exposure caused by the accident itself.

B.1 National Regulations

European countries completed the integration of the ICRP recommendations into national law over a period of several years, for example, Germany published the radiation-protection regulations in 2001, and France followed in 2003. As progress in this field continues, there is an ongoing process of updating the recommendations, and then the national regulations. The regulations in different countries are not completely identical, but the guidelines are the same all over Europe.

In contrast, other countries, such as the United States of America, have regulations which differ substantially from those recommended by the ICRP. For example, the annual whole-body dose limit for workers exposed to ionising radiation in the US is 50 mSv compared to 20 mSv in European countries. Other differences are that in the US the old radiation units, rad, rem and Curie, are still in use (1 Sv = 100 rem, 1 Gy = 100 rad, 1 Curie = 37 billion Becquerels). In most other countries, the ICRP recommendations have been adopted for the national regulations. In the following, some detailed features of the European regulations are presented, which should cover the important points for most countries, leaving out the minor modifications for specific countries.

The object of the European regulations is that, while working with radiation sources or radioactive substances:

- The maximum permissible doses compatible with sufficient safety are respected;
- The maximum levels of exposure and contamination are adhered to; and
- There is appropriate health monitoring for the workers.

The Directive, which sets out the regulations, defines safety standards for exposed workers in the following way:

- The limit on the effective dose is 100 mSv in a consecutive five-year period, subject to a maximum effective dose of 50 mSv in any single year, as mentioned above. In accordance with this, most Member States have defined an annual limit of 20 mSv.
- The annual limit on the equivalent partial-body dose for the lens of the eye is 150 mSv.[2]
- The annual limit on the equivalent partial-body dose for the skin is 500 mSv.
- The annual limit on the equivalent partial-body dose for the hands, forearms, feet and ankles (*extremities*) is 500 mSv.

The reader will remember that doses received on some part of the body can be scaled by the tissue weighting factors (see Table 2.4) to give an equivalent whole-body

---

[2]The ICRP proposed in April 2011 that the annual equivalent dose limit for the lens of the eye, which is the most sensitive tissue in the human body, should be 20 mSv for radiation workers (and should remain 15 mSv for the public).

dose, so, for example, a person receiving 500 mSv on the skin (and no other dose) has received a whole-body equivalent dose of 5 mSv, because the tissue weighting factor for skin is 0.01.

Under exceptional circumstances, like in radiological emergencies, occupational doses for some identified workers may exceed the dose limits above. Workers performing emergency services after accidents, saving lives or protecting large populations from radiation, are allowed to take a single dose of 250 mSv once in their life. However, the full lifetime dose of a radiation worker must not exceed 400 mSv.

The dose limits for apprentices and students aged 18 or over are identical to the ones mentioned. However, the dose limits for apprentices and students aged between 16 and 18 are reduced to 6 mSv per year for the whole-body dose. Correspondingly, the partial-body dose limits for this age group are also lower for the eyes (50 mSv/yr), the skin (150 mSv/yr) and the extremities (150 mSv/yr).

For the general public, the annual dose limit is 1 mSv. In special circumstances, a higher effective dose may be authorised in a single year, provided that the average over five consecutive years does not exceed 1 mSv per year. The annual partial-body dose limit for the eyes for the general public is 15 mSv/yr, and it is 50 mSv/yr for the skin.

The general recommendation is that reasonable steps must be taken to ensure that the exposure of the population as a whole is kept as low as reasonably achievable (the ALARA principle). This last recommendation has been tightened in some countries, such as Germany, to "as low as possible".

To help organisations administer the protection of exposed workers, *controlled areas* and *surveyed areas* (also known as "supervised areas") are defined. Controlled

"Regulation diversity"

© by Claus Grupen

areas must be clearly delineated and access to the area must be restricted to individuals who have received appropriate instruction in radiation-protection standards. For surveyed areas the restrictions are less stringent, and the potential exposures are expected to be smaller. However, radiological surveillance of the working environment must be organised in accordance with the provisions of the standards of radiation protection.

In a surveyed area, a worker might receive up to 6 mSv/yr, and in a controlled area up to 20 mSv/yr. The areas (within controlled areas) where the dose rates are particularly high are called "exclusion areas", and admittance is only permitted for exceptional radiological situations, such as radiation accidents. The qualification to be an exclusion area is different in different jurisdictions, but typically they would be anywhere where the dose rate can be above 3 mSv/h. An example would be the area immediately surrounding the reactor core in a nuclear power station.

For the purposes of monitoring and surveillance, a distinction is made between two categories of exposed workers. A worker must be categorised before starting work in a controlled area. Category-A workers are those who are liable to receive a dose greater than 6 mSv per year (limited to 20 mSv a year), while Category-B workers are those liable to receive a whole-body annual dose over 1 mSv (they may receive a maximum of 6 mSv). The doses actually received must be individually monitored. It also has to be demonstrated that Category-B workers are correctly classified. Should there be an accidental exposure, the relevant doses and their distribution in the body must be assessed.

Any exposures have to be documented for each individual exposed worker, and these records must be retained for the duration of the worker's employment in an exposed environment, and afterwards until the individual has or would have reached the age of 75 years, but in any case not less than 30 years from the termination of the work involving exposure.

As in the ICRP recommendations, the exposure conditions and operational protection of students and apprentices aged 18 or over are equivalent to those of other exposed workers. The lower limit (6 mSv annual whole-body dose) for students and apprentices aged between 16 and 18 means they may only be (up to) Category-B workers, and must be treated as such.

The European Directive closes by asking the Member States of the European Community to bring into force the laws, regulations and administrative provisions necessary to be compliant, before 13th May 2000. No European country was able to implement the Directive within this time limit. The Member States were given some freedom in their implementation of the Directive: they were permitted to adopt dose limits which are stricter than those laid down in the Directive, but not more liberal ones.

One of the most important points in the Directive is the specification of *exemption limits* and *clearance levels* for the quantity or activity concentration of a large number of radioisotopes. Exemption limits are permissible (low) levels of total activity, below which materials do not need to be treated as radioactive. Clearance levels are very similar, but refer to a concentration: either on a surface (in Becquerels per square centimetre) or per unit mass (in Becquerels per kilogram). Usually, if an organisation

Fig. B.1 Design-approved smoke detector with an ^{241}Am source of activity around 30 kBq. ^{241}Am is an α-emitter, and the movement of these charged particles forms part of the electrical circuit. If smoke is present in the detector, some of the particles will be blocked, changing the current. This current change triggers an alarm signal. (Image credit: www.treehugger.com)

is complying with the exemption limits, the competent authority does not need to approve of any disposal or reuse; but in order to treat material complying only with the clearance level as non-radioactive, an application to the competent authority is required. Furthermore, the European Commission has introduced the concept of general clearance levels (in activity per unit mass): default values for materials arising from any practice, any type of material and any pathway of recycling or reuse.[3] The "Guidance on General Clearance Levels for Practices" contains a wealth of information on all conceivable radiation risks and exposures, presented in the form of detailed tables with explanations. It covers man-made radiation equipment as well as natural sources (e.g. radon exposures).

Another useful way to handle radioactive substances in a way that is manifestly compliant with the regulations is to obtain design approval: the design of installations or devices which either contain radioactive substances, or which produce ionising radiation, can be approved if the radiation-protection regulations are respected. To get this approval, the manufacturer or importer of the equipment has to send an application to the competent authority. It is important for the application that the intended use of the installation or device be described in detail. Typical examples of systems with approved designs are smoke alarms containing radioactive sources (see Figure B.1); X-ray sources equipped with an electromagnetic shutter and electrical interlocks; and calibration sources for test measurements.

---

[3]Practical Use of the Concepts of Clearance and Exemption; Part I: Guidance on General Clearance Levels for Practices; Part II: Application of the Concepts of Exemption and Clearance to Natural Radiation Sources; Recommendations of the Group of Experts established under the terms of Article 31 of the European Treaty.

The competent authority will base its decision on design tests, carried out by the authority itself. For this purpose, the manufacturer has to supply design specimens for testing. If the instruments are found to be safe according to the radiation-protection regulations, the design approval will be granted, usually for a period of 10 years. It is possible to extend this period on request, if the instruments are still considered safe.

B.2   Radiation Protection Inside Organisations

In an organisation, the members of staff with responsibility for ensuring the rules are respected have one of two roles: radiation-protection supervisor, or radiation-protection officer (the word "protection" is often dropped from these titles for brevity). There are usually then multiple radiation workers: these members of staff must follow rules and guidance that are given to them, and of course it would be sensible for them to report any possible issues to a responsible member of staff, but they do not have any direct duties with respect to the regulations. An example of the management structure for radiation protection in an organisation is sketched in Figure B.2.

The radiation-protection supervisor is responsible for defining the areas of responsibility of each radiation officer post, including ensuring (if the organisation requires multiple radiation officers) that the responsibilities of the different posts do not overlap. The same supervisor is then responsible for recruiting suitably-qualified people into those positions. The document appointing these officers to these positions is called a radiation-protection directive, and must be communicated to the competent authority. Unlike the radiation officers, the radiation-protection supervisor does not need to be an expert in radiation protection, as the task of understanding and implementing the regulations is delegated to the radiation officers.

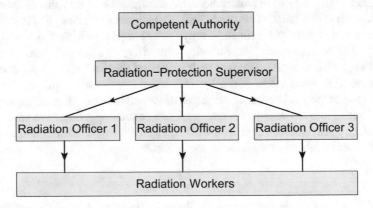

Fig. B.2 Example of a hierarchy for radiation protection within an organisation

The radiation supervisor may only appoint a new radiation officer if there is proof that the officer has relevant and up-to-date qualifications, and that these qualifications are approved by the competent authority. Since the position of a radiation officer comes with important responsibilities, it is necessary to perform a more general check for any facts which could cast doubt on the officer's reliability. The details of the reliability check will vary by jurisdiction.

The radiation officer oversees the handling of the radioactive material in the organisation. He or she must inform the radiation supervisor immediately if there are deficiencies affecting safety. Conflicts may arise where the radiation officer proposes alterations in the implementation of the safety regulations which the radiation supervisor does not approve. However, the radiation officer must not be hindered in any way in performing these duties and should not be put at a disadvantage due to these activities, even if they do not find the full support of the radiation supervisor. The radiation supervisor and the radiation officer must ensure that in the event of danger to life, health or property, appropriate measures are taken immediately. For example, if there is a danger that radioactive substances might be dispersed into the environment, this risk must be kept as low as practicable so that the probability that members of the public receive incorporations is kept at a sufficiently low level. As a further example, steps must be taken to guarantee that accidental criticality of nuclear fuel cannot occur.

There will inevitably be occasional incidents in the practical work of radiation protection about which it is the duty of the radiation officer to inform the competent authority, for example: exceeding the allowed dose limits; loss and acquisition of radioactive material; and deficiencies and problems with safety systems.

The responsibilities of a radiation officer include organising licences, bookkeeping, ensuring there is appropriate training, arranging medical supervision, handling incidents, and dealing with any radioactive waste, as is detailed in the following sections.

B.2.1 Licensing

Usually the handling of radioactive material requires a licence from the competent authority. It is the duty of the radiation officer to arrange the necessary licences for the handling of radioactive material. The radiation officer has to ensure that the radioactive materials in the organisation's possession conform to the materials which are included in the licence. Also the radioactive materials should only be handled by individuals who have had the right training, and are authorised to do so according to the licence. Certainly, all users have to carry the required monitoring equipment such as film badges or electronic dosimeters. A licence will be granted to individuals or companies if there is no evidence casting doubt on the reliability of the applicant.

The transportation of radioactive substances, and especially that of waste containing nuclear fuel, on public traffic routes requires a licence from the competent

© by Claus Grupen

authority (see Section 12.3). Those involved in this transportation have to make sure that possible radiation hazards to the public are kept at a level as low as possible, and they have to make sure that protective measures are taken.

The import or export of radioactive substances or waste containing nuclear fuel also requires a special licence. For imports or exports, it must be ensured that the radioactive material is received only by individuals who are in the possession of the requisite licence for handling them.

One might assume that the licences are only meant for radioactive sources, radioactive substances and nuclear waste, but there are also other applications which require a special licence, including the operation of accelerators for fundamental research or for medicine, and the installation and running of neutron generators and plasma facilities.

If an installation such as a particle accelerator is being modified, or if the operating procedures are being changed in such a way that the aspects of radiation protection are affected, then a new licence is likely to be required.

The radiation officer must also ensure that all radioactive material under the licence is properly secured against unauthorised removal and against theft. Special care has to be taken in the handling of unsealed radioactive sources, because they present an increased risk of incorporation of radioactive material. In addition, the radiation officer has to perform periodic leak tests of sealed sources. If an originally sealed source is found to be leaky, then it takes on the character of an unsealed source. Normally such a leaking source has to be stored away (until appropriate disposal can be organised) because the handling of this leaky source will usually not be covered by the licence.

B.2.2 Bookkeeping

It is very important to keep an up-to-date file with the details of the extraction, production, acquisition, delivery, and handling of radioactive substances. The documentation must contain, in each case, the identity of the radioisotope (e.g. ^{137}Cs), the activity and type of radiation (e.g. 10^6 Bq; β, γ), the physical characteristics (e.g. solid, liquid or sealed), and the date of the acquisition of the radioactive material. The file also has to indicate if the substances are to be used by other individuals or organisations, and then when they are, it has to be ensured that at any moment the whereabouts of the substances are known. A loss of substances has to be immediately reported to the competent authority.

On the other hand, not only losing but also finding radioactive substances might create problems. In any case of discovery, the competent authority must be notified of the find. This includes unintentional acquisitions. It can also happen that a person or installation receives radioactive substances without knowing that they are radioactive. In such cases a measurement of the activity, and a comparison with exemption limits and clearance levels, will clarify whether the competent authority has to be notified.

Radiation doses have to be documented in a radiation passport for the radiation workers. The purpose of this is to prevent the dose limits being exceeded, particularly if radiation workers are active in different companies and locations. The radiation passport shows whether the owner of the passport fulfils the conditions for the task envisaged, for example that he is allowed to enter a radiation-controlled area. The radiation officer of a plant is only permitted to allow activities for people not belonging to the company, on condition that they provide an uninterrupted, complete radiation passport.

Labelling is very important, not only on the radioactive materials themselves, but also on any containers and on any rooms in which they are being stored and used. Similarly, devices which produce ionising radiation (such as accelerators and X-ray tubes) must be appropriately labelled. Also the different radiation areas (exclusion areas, controlled areas, surveyed areas) and contaminated areas have to be marked as such.

The labelling must contain information about the radioisotope, its activity and the responsible radiation officer. Labels which are no longer valid have to be removed.

B.2.3   Instruction and Training

Individuals authorised to enter restricted areas must receive appropriate instruction before entering these areas for the first time. The training has to include: normal working procedures; possible dangers; possible radiation exposures; safety and protective measures; and the licence which governs their work. Those handling radioactive material or using ionising radiation outside restricted areas must also receive such training, if the work requires a licence. The contents of the training must be documented and signed by the individuals who have received it.

B.2.4   Medical Supervision

The medical surveillance of Category-A workers is the responsibility of approved medical practitioners or approved occupational health services. This medical surveillance includes a medical examination prior to employment or classification as a Category-A worker, and further medical examinations at least once a year.

The aim of the medical examination is not primarily to diagnose radiation damage but rather to ensure that these workers are still able to fulfil their job in restricted areas. For example, if the respiration of a particular worker is such that he or she can no longer wear a breathing apparatus, then that worker should not be permitted to work with unsealed sources which give off gases (as it is compulsory to wear a breathing apparatus for that work).

It is even possible to exclude certain workers from working in a radiation-exposed area – this is often done, for example, for pregnant women. The restriction or reduction of tasks for individuals in controlled areas can also be the consequence of exceeding the allowed annual whole-body dose. The purpose of these restrictive procedures is to limit further radiation exposures so that the dose over a period of several years stays within the allowed dose limits.

B.2.5 Handling of Incidents

An important aspect in the field of practical radiation protection is the handling of radiation incidents. It is necessary to distinguish between incidents of different severities using four categories: minor incidents, accidents, serious accidents and emergency situations. A situation does not have to be very severe to be called an accident: if ever the normal running of a plant has to be interrupted for safety reasons, this is termed an accident. When an installation or plant is being designed, it is important to ensure that accidents are foreseen and planned for, and so if they do occur, then accidental doses will not exceed the values given in the relevant radiation-protection regulations. In the case of an incident, the permitted annual dose for radiation workers might be exceeded. This is allowed when a higher "accidental dose limit" (50 mSv in the EU) is respected, and if in addition, the worker does not exceed the normal annual limit on average over five years.

Serious accidents and emergency situations are quite rare. However, any radiation incident must be handled with great care to ensure maximum possible safety for the radiation workers and the general public. When there is an incident in a radiation area, there is a very clear priority order for tasks which must be followed, namely: rescue first, then alert, then secure the area, and finally inform the competent authority.

There are special classifications of areas in technical installations for use in case of fire. These (three) different categories are related both to the total amount of activity, and to any specific dangers (in terms of open or sealed sources), in the various parts of the installation (see Section 12.7). These areas have to be properly labelled for the information of the fire brigade.

B.2.6 Disposing of Radioactive Waste

A radiation officer in an organisation may be responsible for the disposal of radioactive waste, if it is on a sufficiently small scale, such as that coming out from a hospital practising nuclear medicine. Clearly, larger-scale coordination is required where the waste produced is on a larger scale, as it is from a nuclear power plant.

Procedures to dispose of radioactive waste are defined in the radiation-protection regulations. There are two compulsory labelling requirements for disposal of radioactive waste:

- the items themselves must be labelled, including identifying the radioisotope;
- the waste container must be labelled with detailed information on the amount of radioactive waste, total activity, dose rate at a distance of one metre, date of storage, and any other relevant quantities.[4]

---

[4]If the container material itself is significantly radioactive (see Section 12.1), this must also be made clear on the label.

It is important to note that it is not permitted to subdivide the radioactive waste into quantities which fall below the exemption limits or clearance levels (see Section B.1). Also the dilution and dispersal of radioactive waste, or any other radioactive material, into the environment is forbidden.

Summary

Most countries follow the recommendations of the International Commission on Radiological Protection in defining and implementing safety standards for radiation protection. In most cases, the dose limit for radiation workers is 20 mSv per year, and the corresponding limit for the general public is 1 mSv per year. As a rule the ALARA principle, to keep the doses as low as reasonably achievable, can be considered as a sound guideline for radiation protection. It is important to keep in mind that additional doses must be compared to the background radiation of about 2.5 mSv per year present nearly everywhere on Earth. The radiation-protection supervisor of an installation defines the area of competence for each radiation officer. The radiation officer, who must be appropriately qualified, typically has responsibilities including organising licences, bookkeeping, arranging training and medical supervision, handling any incidents, and dealing with waste.

Appendix C
Periodic Table of Elements

Group

Ia	IIa	IIIb	IVb	Vb	VIb	VIIb	VIIIb	VIIIb	VIIIb	Ib	IIb	IIIa	IVa	Va	VIa	VIIa	VIIIa
1 **H** Hydrogen 1.01																	2 **He** Helium 4.00
3 **Li** Lithium 6.94	4 **Be** Beryllium 9.01											5 **B** Boron 10.81	6 **C** Carbon 12.01	7 **N** Nitrogen 14.01	8 **O** Oxygen 16.00	9 **F** Fluorine 19.00	10 **Ne** Neon 20.18
11 **Na** Sodium 22.99	12 **Mg** Magnesium 24.31											13 **Al** Aluminium 26.98	14 **Si** Silicon 28.09	15 **P** Phosphorus 30.97	16 **S** Sulphur 32.07	17 **Cl** Chlorine 35.45	18 **Ar** Argon 39.95
19 **K** Potassium 39.10	20 **Ca** Calcium 40.08	21 **Sc** Scandium 44.96	22 **Ti** Titanium 47.87	23 **V** Vanadium 50.94	24 **Cr** Chromium 52.00	25 **Mn** Manganese 54.94	26 **Fe** Iron 55.85	27 **Co** Cobalt 58.93	28 **Ni** Nickel 58.69	29 **Cu** Copper 63.55	30 **Zn** Zinc 65.39	31 **Ga** Gallium 69.72	32 **Ge** Germanium 72.64	33 **As** Arsenic 74.92	34 **Se** Selenium 78.96	35 **Br** Bromine 79.90	36 **Kr** Krypton 83.80
37 **Rb** Rubidium 85.47	38 **Sr** Strontium 87.62	39 **Y** Yttrium 88.91	40 **Zr** Zirconium 91.22	41 **Nb** Niobium 92.91	42 **Mo** Molybdenum 95.94	43 **Tc** Technetium 97.91	44 **Ru** Ruthenium 101.07	45 **Rh** Rhodium 102.91	46 **Pd** Palladium 106.42	47 **Ag** Silver 107.87	48 **Cd** Cadmium 112.41	49 **In** Indium 114.82	50 **Sn** Tin 118.71	51 **Sb** Antimony 121.76	52 **Te** Tellurium 127.60	53 **I** Iodine 126.90	54 **Xe** Xenon 131.29
55 **Cs** Caesium 132.91	56 **Ba** Barium 137.33	57-71 Lanthanides	72 **Hf** Hafnium 178.49	73 **Ta** Tantalum 180.95	74 **W** Tungsten 183.84	75 **Re** Rhenium 186.21	76 **Os** Osmium 190.23	77 **Ir** Iridium 192.22	78 **Pt** Platinum 195.08	79 **Au** Gold 196.97	80 **Hg** Mercury 200.59	81 **Tl** Thallium 204.38	82 **Pb** Lead 207.20	83 **Bi** Bismuth 208.98	84 **Po** Polonium 208.98	85 **At** Astatine 209.99	86 **Rn** Radon 222.02
87 **Fr** Francium 223.02	88 **Ra** Radium 226.03	89-103 Actinides	104 **Rf** Rutherfordium 261.11	105 **Db** Dubnium 262.11	106 **Sg** Seaborgium 263.12	107 **Bh** Bohrium 262.12	108 **Hs** Hassium 277.15	109 **Mt** Meitnerium 268.14	110 **Ds** Darmstadtium 271.15	111 **Rg** Roentgenium 272.15	112 **Cn** Copernicium (285)	113 **Uut** Ununtrium (286)	114 **Fl** Flerovium (289)	115 **Uup** Ununpentium (289)	116 **Lv** Livermorium (293)	117 **Uus** Ununseptium (294)	118 **Uuo** Ununoctium (294)

Lanthanide series	57 **La** Lanthanum 138.91	58 **Ce** Cerium 140.12	59 **Pr** Praseodymium 140.91	60 **Nd** Neodymium 144.24	61 **Pm** Promethium 144.91	62 **Sm** Samarium 150.36	63 **Eu** Europium 151.96	64 **Gd** Gadolinium 157.25	65 **Tb** Terbium 158.93	66 **Dy** Dysprosium 162.50	67 **Ho** Holmium 164.93	68 **Er** Erbium 167.26	69 **Tm** Thulium 168.93	70 **Yb** Ytterbium 173.04	71 **Lu** Lutetium 174.97
Actinide series	89 **Ac** Actinium 227.03	90 **Th** Thorium 232.04	91 **Pa** Protactinium 231.04	92 **U** Uranium 238.03	93 **Np** Neptunium 237.05	94 **Pu** Plutonium 244.06	95 **Am** Americium 243.06	96 **Cm** Curium 247.07	97 **Bk** Berkelium 247.07	98 **Cf** Californium 251.08	99 **Es** Einsteinium 252.08	100 **Fm** Fermium 257.09	101 **Md** Mendelevium 258.10	102 **No** Nobelium 259.10	103 **Lr** Lawrencium 262.11

For each element the proton number (top left) and atomic mass (bottom) is given. The atomic mass is weighted by the isotopic abundance in the Earth's crust. Bracketed atomic masses are estimates.

© Springer International Publishing Switzerland 2016
C. Grupen and M. Rodgers, *Radioactivity and Radiation*,
DOI 10.1007/978-3-319-42330-2

Appendix D
Further Reading

James E. Martin
Physics for Radiation Protection
Wiley-VCH (2013)

Glenn F. Knoll
Radiation Detection and Measurement
John Wiley (2010)

Wade Allison
Radiation and Reason: The Impact of Science on a Culture of Fear
Wade Allison Publishing (2009)

Wade Allison
Nuclear is for Life: A Cultural Revolution
Wade Allison Publishing (2015)

Sören Mattsson and Christoph Hoeschen
Radiation Protection in Nuclear Medicine
Springer (2013)

Claus Grupen
Introduction to Radiation Protection
Springer (2010)

© Springer International Publishing Switzerland 2016
C. Grupen and M. Rodgers, *Radioactivity and Radiation*,
DOI 10.1007/978-3-319-42330-2

Glossary

> *A good notation has a subtlety and suggestiveness which at times make it almost seem like a live teacher.*
>
> Bertrand Russell 1872–1970

Absorber

Any material used to shield against ionising radiation. X rays and γ rays are best absorbed with materials with a high proton number (e.g. lead or tungsten). The absorption of neutrons is best done with substances containing a lot of hydrogen in their molecules, such as water or paraffin. Neutrons can also be captured in boron or cadmium. α rays are of very short range and can be stopped even with thin foils. β rays should be absorbed with low-proton-number materials (1 cm aluminium or 2 cm plastic) to avoid bremsstrahlung.

Accelerator

A device which uses electric and magnetic fields to speed up electrically charged particles to high velocities, and to focus them. Electron linear accelerators and ion accelerators are used in tumour therapy.

Activation

Material which is struck by neutrons, protons or α particles can become radioactive, although this is only common for neutron bombardment. This is called activation. Although activation with γ and β rays is in principle possible, it is so unlikely that the effect can be ignored.

Activity

The number of nuclear transformations per second. The unit of activity is the Becquerel (Bq). 1 Bq is 1 decay per second. The old unit of the activity (still in use in some countries) was the Curie; 1 Curie corresponds to the activity of 1 gram of radium-226. 1 Ci $= 3.7 \times 10^{10}$ Bq, 1 Bq $= 27$ pCi.

© Springer International Publishing Switzerland 2016
C. Grupen and M. Rodgers, *Radioactivity and Radiation*,
DOI 10.1007/978-3-319-42330-2

"I'd prefer to stick to the old activity unit: 'micro-Curie' sounds so much better than 'Mega-Becquerel'."

Acute Radiation Syndrome
See *radiation sickness*.

Air Pollution
Toxic or radioactive gases and aerosols introduced into the atmosphere by nuclear facilities and also by coal-fired power plants. In normal operation, a coal-fired power plant releases more radioactivity into the environment than a nuclear power plant.

ALARA Principle
It is recommended to keep the radiation exposure 'as low as reasonably achievable' when handling radioactive material or working in nuclear facilities. Some national regulations have the stronger requirement to keep the radiation exposure 'as low as possible'. However, it does not make practical sense to reduce the radiation level well below the dose received from the natural environment.

α Decay
The emission of an α (pronounced "alpha") particle, which is the same as a helium nucleus, from a heavy nucleus. In α decay the atomic mass changes by 4 units, and the proton number by 2 units, e.g. $^{238}_{92}\text{U} \rightarrow {}^{234}_{90}\text{Th} + \alpha$.

Annihilation

The destruction of a particle and its antiparticle as a result of a collision between them: for example, an electron (e^-) and a positron (e^+) may interact resulting in the destruction of both particles and the creation of two (or three) photons $(e^+e^- \rightarrow \gamma\gamma)$. Annihilation radiation is used in medical diagnosis, specifically in Positron-Emission Tomography (PET).

Annual Incorporation

Also called *annual intake*. The amount of radioactive material taken into the body of an adult human as a result of inhalation or ingestion in a year. The national radiation-protection regulations limit the intake of any radioactive substances discharged from nuclear facilities. The limits vary from country to country. In Europe the limit is 0.3 mSv per year.

Antimatter

For each charged particle (e.g. e^-) there is a particle with opposite charge (e^+), but otherwise identical properties. If the first particle is normal matter, the one with opposite charge is called antimatter.

Atom

The smallest possible unit of an element. It consists of an atomic nucleus and a number of electrons orbiting. For a neutral atom, the number of electrons circling will be equal to the number of protons in the nucleus. If the numbers differ, there is a net overall charge, and it is called an ion. A typical atomic size is 10^{-10} m, and a typical mass 10^{-25} kg.

Atomic Energy

See *nuclear energy*.

Atomic Mass

The atomic mass, or "atomic mass number" of a nucleus is the total number of nucleons it contains. It is equal to the sum of the number of protons and the number of neutrons. It is usually written A.

Atomic Nucleus

The positively charged nucleus of an atom. The nucleus consists of a number of protons (Z) and a number of neutrons (N). These constituents, collectively called nucleons, are bound by strong interactions, which are stronger than the repulsive force acting between the positively charged protons. The atomic mass of a nucleus, written A, is the total number of nucleons in a nucleus, so it is given by the sum of the proton number and the neutron number $(A = Z + N)$. Nuclei typically have a diameter of several femtometres $(10^{-15}$ m$)$.

Atomic Number

Another term for the proton number. This term is not preferred in this book, because of potential confusion caused by the similarity of the term to "atomic mass".

Attenuation

As a beam of particles moves through matter, its intensity will be reduced by inter-actions with that matter. This reduction in intensity is called attenuation. Denser materials will usually attenuate more strongly. This term is most often used with respect to photons, but it is also true to say that a beam of α or β particles (or of any other type) is attenuated on passing through matter.

Background Radiation

Normal ambient radiation. This comes from cosmic rays and from radioactivity in soil, water and air. Clearly, this radiation often interferes with low-count-rate measurements. A typical annual dose due to background radiation for a human is 2.5 mSv.

Band Theory

A model of the energy levels in a solid. It applies to crystalline solids, which most inorganic solids are. It says that there are certain energy levels (bands) in which electrons can sit. The lowest-energy bands will be fully filled, and the highest-energy ones will be empty. If there is a band which is partially filled, the material will be a conductor. If the lowest energy band which is not completely full (the *conduction band*) is actually empty, the material will be an insulator. If, in an insulator, it is easy to push electrons into the conduction band using heat or collisions with incoming particles, the material is a semiconductor.

Becquerel

The most widely used unit of activity, often written Bq. 1 Bq is one decay per second; $1\,\text{Bq} = 27\,\text{pCi}$, $1\,\text{Ci} = 3.7 \times 10^{10}\,\text{Bq}$.

β Burns

β particles, because of their short range in tissue, cause damage mostly near the skin's surface. For high exposures, this damage can appear in the form of burns. The extent of these β burns depends on the intensity of the radiation and the length of the exposure time. For very high doses, a similar symptom can also be caused by γ rays (a γ burn).

β Decay

β (pronounced "beta") decay is a type of radioactive decay in which a nucleus transforms into another one, differing in charge by 1 unit ($+1$ or -1). There are three types of β decay: β^- emission, β^+ emission, and electron capture ($p + e^- \rightarrow n + \nu_e$).

β Radiation

Electron emission from an atomic nucleus, after a neutron transforms into a proton ($n \rightarrow p + e^- + \bar{\nu}_e$). The energy released by this decay is always the same for any particular radioisotope, but this energy is shared, in varying proportions, between the proton (p), electron (e^-) and the antineutrino ($\bar{\nu}_e$). Therefore the electron energy spectrum is continuous up to a maximum energy. The most probable electron energy is about one third of the maximum energy. Positron emission ($p \rightarrow n + e^+ + \nu_e$) is also classified as β radiation.

Binding Energy

The energy input which would be required to decompose an atomic nucleus into its constituent nucleons, i.e. to separate the protons and neutrons of a nucleus from each other completely. Typically, this energy is 7–8 MeV per nucleon. It is highest for intermediately heavy nuclei, peaking at iron-56. The binding energy (per nucleon) is lower for both very heavy and very light nuclei. This means that energy is released when light nuclei are fused together, and energy is also released when heavy nuclei fission.

Biological Half-Life

The time required for the amount of a radioactive substance in a biological system to be reduced by a factor of two by metabolic processes. The biological half-life does not include radioactive decay.

Boiling-Water Reactor

A power-plant design in which the water flows through the nuclear core where it is heated and allowed to boil. The steam produced drives a turbine, the motion of which produces an electric current in a generator.

Bone Seeker

See *critical organ*.

Brachytherapy

A form of radiotherapy in which a radioactive source is placed in direct contact with a tumour, either inside the patient's body, or on the skin.

Bragg Peak

When a heavy charged particle, such as a proton or a carbon-12 nucleus, moves through matter, it gives up most of its energy at the end of its range. This means the maximum amount of energy deposited, and therefore the maximum amount of damage, can be targeted at a precise depth in the body. This feature is exploited in hadron therapy.

Breeder Reactor

A nuclear reactor in which, alongside the generation of electricity, transuranic elements are also created ("bred") by causing neutrons to attach to uranium. β^- decay causes elements of higher proton number to appear. A breeder reactor can generate more fissile material than it consumes. Weapons-grade plutonium originates in this kind of reactor. If the neutrons which are being made to attach are high-energy (i.e. not completely moderated), the reactor is called a *fast breeder*.

Bremsstrahlung

The emission of electromagnetic radiation during the deflection or deceleration of charged particles in the electric field of a nucleus. Although all charged particles emit this radiation to some extent on passing through matter, for practical purposes, this only applies to electrons. The word is a German one, meaning "breaking radiation".

Carbon Dating

A dating method for archeological objects, which works by measuring their remnant carbon-14 activity. ^{14}C is a natural, radioactive form of carbon which is quite common on Earth because it is constantly being created by cosmic rays. It has a half-life of 5730 years. Living beings are constantly taking on new carbon from the air or from food, and so the radioactive ^{14}C is replenished as fast as it decays. This means that all living things have a constant quantity of ^{14}C in every kilogram of their bodies, which leads to an activity of about 50 Bq per kg. After the death of the organism, the replenisment stops, and the ^{14}C decays over time. This means that a measurement of the amount of ^{14}C remaining in an organism can tell us how long ago it died, and therefore the age of the object (for example, a wooden axe handle). See also *dating*.

Category-A Worker

A radiation worker who might receive a whole-body dose of more than 6 mSv/yr. The annual limit of 20 mSv/yr must of course still be respected.

Category-B Worker

A radiation worker who might receive a whole-body dose of more than 1 mSv/yr, but should not get more than 6 mSv/yr.

Chain Reaction

A sequence of reactions where a reaction product causes additional reactions to take place. An example is the fission chain reaction initiated by neutrons in a nuclear reactor.

Cherenkov Light

The speed of light in any material is slower than that in the vacuum. This means that a fast-moving particle can be moving more quickly than the speed of light in the medium around it (it cannot, of course, be travelling faster than the speed of light in a vacuum, as this is an ultimate limit). If, for example, an electron travelling through water does exceed the speed of light in that water, it gives out a characteristic blue light, called Cherenkov light. The mechanism is similar to that for the sonic boom given out by a supersonic jet: the electron catches up with the light that it itself has emitted forwards, and the light piles up, giving a flash.

Chernobyl Accident

The most severe reactor accident in history, which happened in Ukraine in 1986, and caused global radioactive pollution. The exposures of the population near to the plant were very high. At greater distances from the plant (a few hundred kilometres or more), the exposures were small but measurable.

Chronic Exposure

Repeated exposure to radiation over an extended period of time. This can cause long-term health effects, but immediate impacts are not expected. The opposite is acute exposure over a very short time, see *radiation sickness*.

Clearance

If the concentration of the activity in a particular piece of radioactive material falls below a certain level (clearance level), then with the permission of the competent authority, it may be exempted from the regulations of radiation protection, and can then be used or reused freely. The clearance levels are defined in national regulations for each radioisotope separately. They give, for example, limits on the activity per kilogram, and on surface contaminations. The clearance level is a very similar concept to the exemption limit, but an exemption limit refers to a total amount of radioactivity, rather than a concentration. If the material complies only with the clearance level, permission from the competent authority is needed for its release.

Competent Authority

In each jurisdiction, there is an authority which supervises radiation protection. It is this authority, called the competent authority, which issues licences, and to which organisations using radioactive substances must report. The competent authority is usually either a specific radiation protection commission, or a health authority. For example: in the USA, the United States Nuclear Regulatory Commission is the competent authority; in the UK it is the Health and Safety Executive; in Switzerland it is the Federal Office of Public Health; in Poland it is the National Atomic Energy Agency; and in Germany it is a different authority in each of the 16 Länder.

Compton Effect

The scattering of energetic photons off electrons inside a material, resulting in the photon leaving with less energy, and the electron with more. In the Compton effect, the binding of the electron to the nucleus is ignored, which is an accurate approximation if the energy of the collision is much higher than the energy binding the electron into the atom.

Computed Tomography (CT)

An X-ray technique for examining cross-sectional areas of the human body. Because it involves the taking of multiple X-ray images, the dose for whole-body CT scans is somewhat higher than for other nuclear medical procedures (perhaps 12 mSv), so adequate medical justification is needed, particularly if these scans are to be repeated.

Conduction Band

The *band theory* describes the energy levels in a solid, saying that there are bands in which electrons can sit. The highest-energy bands will be empty. Of the bands which are either empty or partially filled, the lowest-energy one is called the conduction band. The band directly below it in energy will be the *valence band*. If the conduction band is partially filled, the material will be a conductor. If it is empty, the material will be an insulator. Radiation will sometimes promote electrons from the valence band to the conduction band.

Containment

The prevention of the escape of radioactive materials from a nuclear reactor by appropriate provisions, particularly after accidents or failures. The reactor in the Three-Mile-Island accident near Harrisburg, Pennsylvania (USA) had good containment and after a meltdown, all the fuel remained within the site. The Chernobyl reactor had no containment with disastrous effects on the environment after the core meltdown.

Contamination

Pollution, in this context by radioactive substances. The term contamination is often used as a shorthand specifically for the contamination of a particular surface, say, of clothing or of skin.

Control Rods

The components of a reactor which are used to manage the reactor reactivity. Control rods contain strong neutron absorbers, like boron or cadmium. When they are inserted further into the reactor core, the reaction rate decreases.

Controlled Area

A radiation area in which individuals (e.g. Category-A workers) might receive a whole-body dose of more than 6 mSv/yr. The maximum annual dose is 20 mSv/yr. For workers of Category B who work in controlled areas, the maximum annual dose is 6 mSv/yr (these limits refer to European regulations).

Core Meltdown

If the normal and the emergency core cooling systems fail simultaneously in a nuclear reactor, the residual heat created by the decay of the fission products heats up the reactor core, until it melts (this happened in the Three Mile Island, Chernobyl and Fukushima accidents).

Cosmic Rays

Energetic particles originating from the Sun, our galaxy (the Milky Way), and beyond, that hit the Earth's atmosphere. About 85 % of all incoming cosmic-ray particles are

"Somebody has contaminated our instruction manual.
Now we can only consult it in our glove box!"

protons, about 12 % are helium nuclei (α particles), and 3 % are nuclei heavier than helium. In addition, there are electrons (about 1 %). Initial (or "primary") cosmic rays generate further particles, called secondary cosmic rays, by interactions in the atmosphere.

Critical Mass
The smallest amount of fissile material needed for a sustained nuclear chain reaction.

Critical Organ
The organ of the body which is most sensitive to radiation damage by a specific incorporated radioactive nuclide or type of external radiation. For example, ^{131}I accumulates in the thyroid gland, the only gland to use iodine. By contrast, ^{90}Sr accumulates in the bones, as it is chemically similar to calcium, a major component of bone (it is therefore called a "bone seeker").

Criticality
A critical amount of fissile material is needed for a sustained nuclear chain reaction. For stable operation the neutron amplification factor must be 1, i.e. the neutrons from each fission should cause exactly one further fission on average. In that case the reactor is called critical. A subcritical reactor, with a neutron amplification factor of less than one, will tend to shut down and produce no power. If a reactor can remain supercritical, i.e. have neutron amplification factor greater than one, for a period of time, the reaction rate will increase continuously, which is potentially dangerous.

Curie (Ci)
An old activity unit, still in use in some countries. One Curie corresponds to the activity of 3.7×10^{10} Bq. It is the activity of one gram of radium-226.

Dating

Certain events can be dated by comparing the amount of a particular radioisotope remaining in a sample, particularly if there is a good non-radioactive element with which a ratio can be measured. For example, a mineral might contain the long-lived isotope potassium-40 (^{40}K, half-life 1.3 billion years). This decays into argon-40. Because argon is a gas, it would escape from liquid rock, so the amount of ^{40}Ar trapped in a solid rock, when compared with the amount of remaining ^{40}K, can tell a geologist how long it is since the rock solidified. The long-lived isotopes ^{232}Th and ^{238}U can be used in a similar way for geological or cosmological dating, and ^{14}C can be used to determine the time since an organism died (see *carbon dating*).

Decay

A process in which an atomic nucleus spontaneously emits one or more α, β or γ particles or neutrons. In doing so, the nucleus changes in type.

Decay Chain

When an unstable nucleus decays, it might not decay directly into a stable isotope. Instead, the first nucleus's decay products might be unstable, and decay themselves. In this way, a series of isotopes are linked in a chain by successive radioactive decays. The chain ends when a stable nuclide is reached. For example: ^{220}Rn→^{216}Po →^{212}Pb →^{212}Bi → ^{208}Tl → ^{208}Pb (the αs and βs which are emitted are not shown here). The last nucleus, lead-208, is stable.

Decay Constant

This constant, usually written λ (pronounced "lambda"), is a convenient number, which is specific to each radioactive isotope. If it is multiplied by the number of nuclei of that isotope, the result is the activity from that isotope. The decay constant is larger for nuclei with a smaller half-life (in fact, they are related by $\lambda = \ln(2)/T_{1/2} \approx 0.7/T_{1/2}$), because nuclei with shorter half-lives have high activities and decay more rapidly. The number of atoms of an isotope remaining in a sample at a particular time (N) can be calculated using the decay constant in the formula $N = N_0 e^{-\lambda t}$, where N_0 is the number of those nuclei at the start of the observation time, and t is the time since there were N_0 nuclei. The decay constant λ is related to the lifetime τ by $\lambda = 1/\tau$.

Decay Law

The law describing the decrease in the number of nuclei over time by radioactive decay. The number of nuclei of a radioisotope remaining falls exponentially: i.e. it halves after a certain time (called the half-life), and falls to a quarter of the original number after another half-life, etc.. The rate at which the number of nuclei decreases is related to the number of nuclei (of that type) which are present. It can be written mathematically as $N = N_0 e^{-\ln(2)t/T_{1/2}}$, where $T_{1/2}$ is the half-life, N is the number of remaining nuclei, N_0 is the number of those nuclei at the start of the observation time, t is the time since the start of the period, and $\ln(2)$ is a constant ($\ln(2) \approx 0.7$).

Decontamination

The elimination or reduction of a contamination.

Decorporation
The removal of radioactive material from the human body by excretion (defecation, urination, sweating, etc.) or vomiting.

Delayed Neutron
In a fission reactor, some neutrons are emitted not immediately during the splitting of the heavy nucleus, but later, from one of the fission products. These neutrons coming from the fission products are called delayed neutrons. They are typically emitted 0.1 s after the fission.

Depleted Uranium
Uranium with a fraction of ^{235}U smaller than the 0.7 % found in natural uranium, so that it is almost all ^{238}U (which has a longer half-life). This is produced as a by-product of uranium enrichment, i.e. it is the part removed from natural uranium to produce enriched uranium. Another source is uranium which has already been used as fuel, and so has lost some ^{235}U via fission (*spent uranium*).

Deuterium
A stable isotope of hydrogen, which has one neutron in addition to the proton in the nucleus, ^2H. It is also called heavy hydrogen. It can be written with the symbol D, so D_2O is water with two deuterium atoms instead of normal hydrogen. It is common in nuclear physics to write a nucleus of deuterium (called a *deuteron*) as d.

Deuterium-Tritium Fusion
The fusion process which is being researched on Earth at the moment (there are, so far, no commercially operational power plants). In this process, one nucleus each of deuterium (^2H) and tritium (^3H) are fused to make a nucleus of helium-4 and a neutron. Although this process requires enormous temperatures and pressures, the requirements are less severe than they would be if hydrogen fusion were being attempted.

Dirty Bomb
A weapon that combines some amount of radioactive material with conventional explosives. The main purpose of the weapon is to contaminate the area with radioactive substances and to cause panic among the population. Dirty bombs should not be confused with nuclear weapons, as they have no more explosive power than a conventional explosive.

Disposal
The removal of radioactive waste for final storage in a dump or disposal site.

Dose
A general term for the extent of an exposure to radiation. In this book, unless otherwise stated, "dose" refers to an effective whole-body dose. This is measured in Sieverts.

Dose Equivalent
See *equivalent dose*.

Dose Rate

The dose absorbed per unit time. Common units for the energy-dose rate are μGy/h and mGy/h. The equivalent-dose rate and effective-dose rate are measured in μSv/h or mSv/h.

$E = mc^2$

Perhaps the most famous equation in all of science. It indicates that energy can be turned into mass, and vice versa, under certain conditions. There is a conversion of (a tiny fraction of) mass into energy during fusion and fission, as the products are lighter than the inputs. There is a conversion of energy into mass, for example, in *pair production*.

Effective Dose

The measure of the dose which takes into account the type of radiation and the sensitivity of particular tissues and organs to that radiation. When the term "dose" is used in general, this is the quantity which is meant. It is also called the *effective whole-body dose*, or the *effective dose equivalent*.

Effective Half-Life

The time required for the concentration of a radioisotope in a biological system to be reduced by a factor of two. The effective half-life ($T_{1/2}^{\text{eff}}$) includes both biological excretion and radioactive decay. It is always smaller than both the biological half-life ($T_{1/2}^{\text{bio}}$) and the physical half-life ($T_{1/2}^{\text{phys}}$), to which it is related by $T_{1/2}^{\text{eff}} = \frac{T_{1/2}^{\text{phys}} \, T_{1/2}^{\text{bio}}}{T_{1/2}^{\text{phys}} + T_{1/2}^{\text{bio}}}$.

Electromagnetism

A force acting between charged and magnetic objects. The electric field causes like charges to repel, and unlike charges to attract. Magnets create a force on moving charges, and on other magnetic or magnetisable objects. These two different effects have been shown to be part of a larger framework, so the words are combined in electromagnetism.

Electron

The negatively-charged component of an atom. Electrons orbit the atomic nucleus. An electron is about 2000 times lighter than a proton or a neutron (9.1×10^{-31} kg).

Electron-Volt (eV)

A unit of energy frequently used in particle and nuclear physics. Using $E = mc^2$, this can also be used as a unit of mass. One eV is the energy gained by a singly charged particle if accelerated in a potential of one volt (1.6×10^{-19} J). Typical energies of α, β and γ rays in the field of radiation protection are in the MeV range. X rays used in medical imaging have energies around 10–100 keV.

Energy Dose

The amount of absorbed energy per unit mass. The unit of the energy dose is 1 Gray (Gy) = 1 Joule/kg. This type of dose does not take into account the different strengths of biological effect of the different radiation types, nor the different susceptibilities of the different organs of the body.

Energy-Dose Rate

The absorbed energy per unit mass and unit time. Common energy-dose-rate units are μGy/h and mGy/h.

Energy Loss

The reduction in the energy of a moving particle on passing through matter, as it gives up energy by ionisation, excitation or the production of bremsstrahlung.

Enrichment

A technique to change the ratio of isotopes in a material in such a way as to increase the proportion of fissile isotopes. The most common example concerns uranium: in natural uranium, the fissile ^{235}U makes up only 0.7 % of the total, but this must be raised to higher proportions, say 3 %, for power production (and to much higher values, typically 95 %, for weapons). This is usually done by placing the uranium in a centrifuge, and separating the lighter fraction, which will contain more ^{235}U.

Equivalent Dose

The absorbed energy, measured in Joules per kilogram, weighted with the radiation weighting factor, i.e. taking account of the different effects of the different radiation types, but not taking account of the different susceptibilities of the different tissues of the body. The unit of the equivalent dose is the Sievert (Sv); 1 Sv = 100 rem.

Excited State

It happens frequently in atomic and nuclear physics that systems are not in their most
stable configuration. For example, an atom might have one of its electrons in a high
orbit, with space for it to fall into a lower orbit. The electron will eventually fall,
releasing energy in the process, usually in the form of an X ray. The unstable state
is called an excited state, and tends to be marked on diagrams with an asterisk (*).
The resulting stable state is called the *ground state*. The other important example is
that it is possible for nuclei to be arranged in an excited state, in which case they will
decay into the ground state emitting a γ ray.

Exclusion Area

A very high-radiation part of a controlled area, to which access is forbidden for all
people except in extreme circumstances. An example would be the area immediately
surrounding a nuclear reactor. The precise definition depends on jurisdiction, but as
an example, in Germany, anywhere where the dose rate can exceed 3 mSv/h should
be an exclusion area.

Exemption Limit

If the activity of radioactive material falls below a certain level (the exemption limit),
it may be exempted from the regulations of radiation protection and can be handled
freely. The exemption limits are defined in national regulations for each radioisotope
separately. The exemption limit is a very similar concept to the clearance level, but it
refers to a total activity, rather than a concentration. Material with an activity below

the exemption limit does not need the competent authority's approval for any reuse or release.

Fallout
Radioactive substances or dust injected into the biosphere by nuclear-weapon tests or reactor accidents. These can be washed out by rain ('washout') or simply sink to the ground, thereby presenting a radiation risk for the population.

Film Badge
A small badge containing photographic films that is worn by personnel to monitor radiation exposure. Modern film badges with different filters measure the received dose, and can even distinguish γ rays from β particles and neutrons.

Fissile
An isotope is fissile if it undergoes induced fission sufficiently easily that it can be used in a nuclear power station.

Fission
Some heavy nuclei, such as uranium-235, can be caused to split into smaller fragments by neutron impacts (*induced fission*). For some very heavy nuclei, such as fermium-256, *spontaneous fission* can occur: breakup without any need for neutron bombardment.

Fissionable
An isotope that undergoes any spontaneous or induced fission under any circumstances is referred to as fissionable. Most fissionable isotopes are not useful for power stations (they are not fissile), for example because they will only undergo fission rarely, and then only with neutrons of high energy.

Flux
The flux of a particle type is the number passing through a certain area in a certain time. For example, at sea level, there is a flux of muons, originating from cosmic rays, of about one particle per square centimetre per minute.

Fractionated Dose
The delivery of a dose in multiple fractions, separated by a period of time (typically a day), rather than all at once. This technique is used in cancer treatment, as higher radiation doses can be tolerated by tissue if they are applied in this way. This is because the body's repair mechanisms are given a chance to operate between the sub-doses. In treating tumours with γ rays or hadrons, this method takes advantage of the fact that healthy tissue heals more quickly and effectively than cancerous tissue.

Fragmentation
The breakup of a nucleus into smaller fragments in a high-energy collision with another nucleus. This differs from fission because the energy to break up the nucleus comes from the collision itself, rather than any instability of the nucleus.

Fuel Cycle
All the steps in the processing of fissile material as fuel for a reactor. This includes mining, purification, isotopic enrichment, fuel fabrication, storage of irradiated fuel, reprocessing of spent nuclear fuel and disposal.

Fuel Rods
The central elements of a nuclear reactor. They mostly consist of uranium enriched in the isotope ^{235}U.

Fusion Reactor
A nuclear reactor in which the energy generation is based on a fusion process. On Earth, it is deuterium-tritium fusion which is being investigated. At the time of writing, there are no such reactors operating commercially, but there are some operating for research.

γ Rays
Particles of electromagnetic radiation in the MeV range, typically produced in a transition within a nucleus. γ (pronounced "gamma") rays can best be shielded with materials of high proton number (e.g. lead or tungsten).

Geiger-Müller Counter
A radiation detector which measures charged particles and γ rays by their ionisation (amplified in an electric field) in a cylindrical volume. The output signal of a Geiger-Müller counter does not depend on the type of the incident particle, nor on the energy loss in the detector. Also called a GM counter or a Geiger counter.

Gluon
The particle that carries the strong force, in the same way that photons carry the electromagnetic force. Gluons bind quarks together to make nucleons.

Gray (Gy)
The unit of energy dose. It is an energy absorption of 1 Joule per kg. One Gray is 100 rads (radiation absorbed dose). The Gray and the rad are measures for the energy transfer by radiation to an absorber, and unlike the Sievert, do not contain weightings for different biological effects.

Hadron
A particle that can interact using the strong force: i.e. anything made up of quarks. Protons, neutrons, pions and heavy nuclei are all examples of hadrons.

Hadron Therapy
Tumour irradiation with protons or heavy nuclei, taking advantage of the precisely-located energy deposition at the end of their range (see *Bragg peak*). Hadron therapy is particularly useful as it can be targeted very narrowly on a tumour, with relatively little damage to healthy tissue. Proton therapy is a subtype of hadron therapy.

Half-Life

The time after which half of the nuclei of some radioactive substance have decayed. This is characteristic for each radioisotope, and can vary between a fraction of a second and billions of years. When considering biological systems, this can be referred to as the physical half-life (to distinguish it from the biological half-life). The decay of an isotope can be described mathematically using the half-life $T_{1/2}$ in the *decay law* equation $N = N_0\,e^{-\ln(2)t/T_{1/2}}$.

Half-Life, Biological
See *biological half-life*.

Half-Life, Effective
See *effective half-life*.

Half-Life, Physical
See *half-life*.

Heavy Hydrogen
See *deuterium*.

Heavy Water
Water in which most or all of the normal hydrogen is replaced by deuterium. Because it is a good moderator which absorbs very few neutrons, heavy water can enable a pressurised-water system to function with unenriched (natural) uranium.

High-Radiation Area

An area in which the radiation level is (at least sometimes) sufficiently high that measures need to be taken to comply with national-radiation protection regulations. High-radiation areas are divided (in most jurisdictions) into surveyed areas and controlled areas, with the highest-danger parts of controlled areas being exclusion areas. As an example, the ICRP recommendations say that areas in which a worker might receive over 6 mSv per year should be defined as controlled areas.

High-Temperature Reactor

High-temperature fission power reactors operate at very high temperatures (around 1000 °C). This is made possible by the use of special ceramic materials in the reactor core. Most reactors of this type use helium gas as a coolant, instead of water. High-temperature reactors are intrinsically safe, and can also be built in relatively small units (100 MW).

Hormesis

The concept that generally favourable biological responses result from additional exposures to low-level ionising radiation. This is a contentious hypothesis, but has supporting evidence from some studies on non-human animals, and an interesting case in Taiwan, discussed in Section 7.6.

Hot Spot

A location where the radioactivity or radiation is higher than expected. For example, absorption of naturally occurring α-emitters by the tar in smokers' lungs can produce hot spots on lung tissue.

Hydrogen Fusion

The energy production mechanism in the Sun and stars. In this multi-step process, hydrogen (^{1}H) is fused to make helium. This mechanism requires levels of pressure and temperature that we cannot reach with current technology, so it is deuterium-tritium fusion which is being investigated on Earth.

ICRP

The International Commission on Radiological Protection. This body recommends regulations and limits for radiation protection.

Incorporation

The intake of radioactive substances by eating or drinking, breathing, or via lesions.

Induced Fission

The process in which a heavy nucleus is struck by a neutron, and so caused to break into two smaller fragments (and usually some free neutrons). Induced fission is the process driving all currently-operational commercial nuclear power stations.

Inertial fusion

See *laser fusion*.

Ingestion

The intake of radioactive substances by eating or drinking.

Inhalation
The intake of radioactive substances by breathing (including the "breathing" of the skin).

Iodine Tablets
A medicine mitigating against the effects of some nuclear accidents. The element iodine is used in the human body in the thyroid gland. Natural iodine is mostly stable iodine-127. In some nuclear disasters, including Chernobyl, significant quantities of radioactive ^{131}I have been released. If a tablet containing a large amount of stable iodine can be administered to a person before an exposure to ^{131}I, the thyroid will have no requirement for additional iodine, and the radioactive iodine will not be retained by the body: it will be excreted, and not be in the body for long enough to cause significant damage.

Ion
If an atom loses or gains an electron, and so has different numbers of protons and electrons, it is called an ion. Ions, by definition, have a net charge. It is very common for chemical compounds to contain ions.

Ionisation
The separation of atomic electrons from their nuclei by photons (photoionisation) or by charged particles (ionisation by collision). Both of these processes result in the presence of additional positive ions and free electrons in the material or tissue. Although ions themselves are common in chemistry, these additional ions (and free electrons) can damage biological structures. Fortunately, there are biological mechanisms to repair some level of damage.

Ionising Radiation
Radiation, usually α, β, X or γ rays, or neutrons, which can produce ions in interactions with atoms. These interactions may be with the air, with inanimate objects, or with living tissue. If ionising radiation interacts in living tissue, that individual receives a radiation dose.

Isotope
If two nuclei have equal electric charge (i.e. the same number of protons), meaning they represent the same chemical element, but they have different numbers of neutrons, then they are different isotopes of that element. In other words, an isotope is a single combination of a number of protons and a number of neutrons. Different isotopes of a single element will have the same chemical properties, but can have dramatically different radioactive behaviours.

ITER
The International Thermonuclear Experimental Reactor, to be built in Cadarache, France. This test facility will use the tokamak principle to create nuclear fusion. It is in its construction phase at the time of writing. ITER is Latin for "the path", as in "the path to fusion power plants".

JET
The Joint European Torus: a prototype of a fusion reactor based on the tokamak principle, in Culham, UK. Research began in the 1970s, and is still continuing.

Large Hadron Collider (LHC)
The largest particle physics experiment in the world at the time of writing. In it, two beams of protons travel in opposite directions around a 27 km ring on the Swiss-French border, and are brought to collision. The results of the collisions are analysed to find out about the properties of subatomic particles.

Laser
An acronym for Light Amplification by Stimulated Emission of Radiation. Each laser produces light of a single wavelength, often in the visible range, which can easily be focussed, and may be of high intensity. Lasers have a broad range of applications in optical instruments, cutting and welding, scanners, medicine, and elsewhere.

Laser Fusion
A technique to cause nuclear fusion using high-power pulsed lasers. Also called inertial fusion.

Latency Period
The time between irradiation and the first signs of response. The latency period for leukaemia is shorter (15–20 years) than that for the development of other cancers (25–30 years).

LET
Linear Energy Transfer. See *RLET*.

"Large Hadron Collider therapy"

Lethal Dose

The median lethal dose of radiation is defined as the dose corresponding to a death-rate of 50 % within 30 days without medical treatment (in the medical profession, this is referred to as the LD_{50}^{30}). The lethal whole-body dose for humans is around 4 Sv.

Leukaemia

A cancer of the bone marrow. The risk factor for radiation-induced leukaemia as a consequence of a whole-body exposure is 5×10^{-4} per 100 mSv (ICRP figure). Leukaemia has a shorter latency period (perhaps 15–20 years) than other cancers.

Lifetime

The time period in which the number of nuclei of a particular radioisotope has decayed to a level of $1/e$ of its initial value, i.e. to about 37 % of the original number. The lifetime τ (pronounced "tau") is inversely proportional to the decay constant, $\tau = 1/\lambda$, and it is about 45 % longer than the half-life (being related to it according to $\tau = T_{1/2}/\ln(2)$).

Light-Water Reactor

A nuclear-fission reactor that uses ordinary water as both coolant and moderator.

LINAC

Short for linear accelerator. An electrical device for the linear acceleration of sub-atomic particles (usually electrons or protons). Unlike in a synchrotron, there is no magnetic deflection. Electrostatic or magnetic lens elements may be included to ensure that the beam is confined to the centre of the vacuum pipe. LINACs are used in radiology for tumour irradiation and also as injectors for synchrotrons.

LNT Hypothesis

The Linear No-Threshold Hypothesis, namely that the risk factor for stochastic (long term) radiation damage depends linearly on the dose, and that there is no threshold effect, i.e. that there is no level of radiation sufficiently low that its effect on health is zero. This is the standard assumption in calculating the risks related to radioactive doses, and is quite conservative for low doses.

Magnetic Confinement Fusion

A technique to cause nuclear fusion by keeping deuterium-tritium plasma in a closed chamber using magnetic fields.

Magnetic Field

A field, created either by a permanent magnet, or by a current-carrying wire (usually set up as a coil), which causes forces on moving charges and magnetic objects. See also *magnetic flux density*.

Magnetic Flux Density

Technically, this quantity, measured in Tesla, is what is usually meant when a "magnetic field" is discussed. It can also be called *magnetic induction*. In this book, these terms are not used, and "magnetic field" is preferred.

Mammography

A test for breast cancer, by taking an X-ray image of a woman's breasts. Using modern techniques, typical partial-body exposures of the tissue are between 2 and 10 mSv, corresponding to an effective whole-body dose of around 0.5 mSv.

Mass Attenuation Coefficient for Photons

The photon mass attenuation coefficient (written μ_m) is a number describing the absorption and scattering of energetic electromagnetic radiation caused by the photoelectric effect, Compton scattering and pair production. The attenuation of the intensity of the photon beam follows an exponential law: $I = I_0\,e^{-\mu_m x}$, if I_0 is the intensity at $x = 0$ and I the intensity of photons with the initial energy after a thickness x of absorber.

Mass Defect

The deficit between the mass of an isotope (nucleus) and the sum of the masses of its constituent nucleons. Via $E = mc^2$, this is the same as the total binding energy of the nucleus.

Medical Supervision

Appropriate medical supervision is mandatory for radiation-exposed workers of Category A in controlled areas, and also for radiation-exposed workers of Category B if they handle unsealed radioactive sources.

Meltdown

See *core meltdown*.

Metastability

It is common in α and β decay for the resulting nucleus not immediately to be in its lowest-energy configuration. Instead, it is in an *excited state*, and as it reorganises itself into the most stable configuration, it emits the excess energy as a γ ray. Usually, these excited nuclear states have a tiny half-life, under 10^{-11} seconds. However, sometimes the excited nuclear state is more long-lasting, and lasts many seconds. In this case, the state is referred to as "metastable" (i.e. still not stable, but much more stable than would be expected). One such example is ^{99m}Tc, which is produced by the β decay of ^{99}Mo. The "m" marks it as metastable. It has a half-life of 6 hours, and is widely used in medical diagnostics.

Moderation

Neutrons are much more likely to induce fission (in a fissile isotope) if they are travelling slowly, i.e. if they have low energies. Neutrons liberated in nuclear fission have relatively high energies. Before they can induce fission, they have to be slowed down by suitable moderators, such as water or graphite.

Muon

A subatomic particle, similar to an electron, but heavier (by a factor of about 200: 1.8×10^{-28} kg). Muons are unstable, and decay to an electron and two neutrinos. Muons are given the symbol μ (but this has nothing to do with the μ of "micro").

Mutation

The replacement of a single base nucleotide (of DNA) with another nucleotide. Mutations may arise spontaneously, but can also be caused by ionising radiation. Because mutations can be caused in the germ cells (the precursors to eggs and sperm), some effects of radiation can be passed on to the next generation.

Natural Reactor

A high concentration of uranium isotopes in ores can lead to natural fission processes under favourable weather conditions (i.e. the availability of water). It is believed that there are no places on Earth where a natural reactor could be happening today. See *Oklo*.

Neutrino

A type of particle which is neutral and very light. They only interact very weakly with matter, and therefore it is practically impossible to get a sufficiently high dose from neutrinos to make them significant for radiation protection. The current understanding of physics says that there are six types. Firstly, there are three pairings of neutrinos with other particles. These three pairings are with the electron, with the muon, and with a heavier, rarer particle called a tau (written "τ"). Each neutrino is paired to one of these other particles, and a subscript "e", "μ" or "τ" can be written to show which is which. Secondly, there is the distinction between the neutrino and the antineutrino. This is important because in β decay, either an electron and an electron-type antineutrino are given out ($n \rightarrow p + e^- + \bar{\nu}_e$) or a positron and an electron-type neutrino ($p \rightarrow n + e^+ + \nu_e$).

Neutron

The neutral building block of a nucleus. Neutrons have a very slightly higher mass than protons (1.6749×10^{-27} kg). Neutrons are stable inside nuclei (except for in β^--emitters), but a neutron outside a nucleus is unstable, decaying to a proton, electron and antineutrino with a half-life of about ten minutes.

"Did you see it?"
"No nothing."
"Then it was a neutrino!"

Neutron Amplification Factor

In a fission chain reaction, the number of neutrons from each nuclear fission which goes on to cause further fission, on average, is called the neutron amplification factor. For a nuclear reactor to be operating in a stable way (i.e. to be "critical", rather than "subcritical" or "supercritical"), the neutron amplification factor has to be precisely 1. The number of neutrons which comes out of each fission is usually either 2 or 3, so some neutrons must be removed, either by being absorbed by control rods or by being allowed to escape the reaction chamber.

Neutron Source

An apparatus which produces neutrons for use in technical installations, for example, a radium-beryllium source. In that case, α particles from radium decay produce neutrons on interacting with beryllium according to $\alpha + {}^{9}_{4}\text{Be} \rightarrow {}^{12}_{6}\text{C} + n$. Instead of radium, other α-emitters (such as polonium or americium) may be used.

Noble Gases

A group of gases which are chemically very unreactive. The naturally-occurring noble gases are helium, neon, argon, krypton, xenon and radon. Radon has no stable isotopes, but is present in the air because it is constantly being created by the decay of uranium.

Non-Ionising Radiation

Radiation which does not produce ions when interacting with matter. Examples include visible light and radio waves. There are mechanisms (mostly heating) that can cause non-ionising radiation to be harmful to humans, but in general it is less dangerous than ionising radiation.

Nuclear Accident
An unexpected event in a nuclear facility, resulting in an increase in the possibility of radioactive contaminations and exposures to people.

Nuclear Energy
The energy released by nuclear reactions. The process in which a neutron causes an atomic nucleus to split is called (induced) fission. The merging of two light nuclei at very high temperatures is called fusion. An alternative term for nuclear energy is *atomic energy*.

Nuclear Fission
The splitting of a nucleus, normally into two fragments. Frequently, neutrons will be emitted either as part of the fission process (prompt neutrons) or by the products a short time after (delayed neutrons). These neutrons can induce further fissions.

Nuclear Fusion
Light nuclei may be caused to merge, making heavier ones, thereby releasing nuclear binding energy. In fusion experiments conducted so far on Earth, deuterium and tritium are being used, and produce helium-4. The Sun fuses protons, via deuterium and helium-3, also to helium-4, in a process called *hydrogen fusion*.

Nuclear Power Plant
A facility which converts nuclear energy into electrical power. Current nuclear power plants use the energy from fission to produce steam, which drives a turbine and generates electricity.

Nuclear Reactor
A device in which a nuclear chain reaction is initiated, controlled and sustained.

Nuclear Tracers
Substances in which one or more atoms are replaced by radioactive atoms. They can also be called *radioactive tracers*. When a tracer is put into a chemical reaction, more information can be gained about the reaction using observations of the decays of the radioactive atoms. Tracers are particularly useful in biological systems, where chemical reactions are particularly complex.

Nuclear Waste
Radioactive material which is a waste product from nuclear power plants, from the nuclear-fuel cycle or from recycling facilities. This waste has be to disposed of by safe long-term storage.

Nucleon
A collective term for any one of the constituents of an atomic nucleus. A nucleon is either a neutron or a proton.

Oklo
In Oklo, Gabon (Central Africa), an unusual event happened 1.7 billion years ago. The rock there had a high concentration of uranium, and favourable weather conditions (i.e. plentiful rainwater). This allowed a natural nuclear reactor to start. It then ran for

millions of years. A large amount of ^{235}U was processed during the operation of the reactor, so this natural reactor will never restart. In the vicinity of Oklo, over a dozen independent natural reactor sites have been found, but no examples of natural reactors are known elsewhere.

Order of Magnitude
A factor of ten. This is a useful concept in comparing the general scale of two numbers. It is common to round to the nearest order of magnitude when illustrating such a comparison. For example, the mass of the proton $(1.67 \times 10^{-27}\,\mathrm{kg})$ is about three orders of magnitude larger than the mass of the electron $(9.1 \times 10^{-31}\,\mathrm{kg})$, because the factor between them is about 2000 (which we round to $10^3 = 1000$).

Pair Production
As a photon passes through a material, it can interact with the electric field from a nucleus, and in doing so split into a particle-antiparticle pair, normally an electron (e^-) and a positron (e^+). To do this, the incident photon must have sufficient energy to create the pair of particles (via $E = mc^2$).

PET
Positron-emission tomography is a medical imaging technique using the annihilation of positrons. A positron (e^+) is produced by β decay, and annihilates with an electron from an ordinary atom. This causes the emission of two photons with $511\,\mathrm{keV}$ of energy each, which are measured by the PET camera to provide an image of the tissue or organ being examined.

Photoelectric Effect

The process in which an X ray or γ ray strikes an electron in an atom and is absorbed, and the electron receives enough energy to leave the atom, making an ion.

Photon

A particle of light. This can be a particle of visible light, or of any other type of electromagnetic radiation, such as infrared or γ radiation. Although light can also legitimately be thought of as a wave, that view is not discussed in this book.

Pocket Dosimeter

An ionisation chamber built as pen-type dosimeter, on which the received dose can be read immediately.

Pressurised-Water Reactor

A power-plant design in which the water flows through the nuclear core and is heated by it, but it is kept under sufficient pressure that it does not boil. Steam is produced behind a heat exchanger: this then drives a turbine and so produces an electric current.

Prompt Neutron

In a fission reaction, some neutrons are emitted as the fission itself happens, rather than later. These are called prompt neutrons.

Proton

The positively charged building block of a nucleus. The mass of a proton (1.6726×10^{-27} kg) is similar to, but slightly lower than, that of a neutron. Protons are stable particles (as far as current science knows).

Proton Number

The number of protons in a nucleus. Because the number of protons in a nucleus determines the element, giving a proton number identifies an element. The proton number is usually given the symbol Z, and is sometimes called the *atomic number* (not to be confused with the *atomic mass*).

Proton Therapy

Precision radiation therapy for the treatment of well-localised deep-seated tumours with protons. This method relies on the increased energy loss of protons at the end of their range (see *Bragg peak*). The surrounding tissue receives only a much lower exposure. See also *hadron therapy*.

Quality Factor

A measure of the effectiveness of a type of radiation at producing damage in a biological system, now rarely used. It depends on the linear energy transfer (see *RLET*). The quality factor has been replaced by the radiation weighting factor in the conversion of the energy dose to the equivalent dose.

Quark

A basic building block of nucleons: each nucleon is made up of three quarks (bound together by the strong force, carried by gluons). There are two common types of quark: "up" (charge $+\frac{2}{3}$) and "down" (charge $-\frac{1}{3}$). A neutron is composed of two down quarks and one up quark, and a proton of two up quarks and one down quark. There are four further types of quark (strange, charm, bottom and top), all of which are short-lived, and are only produced (for research) in particle accelerators.

rad
Radiation absorbed dose. This is a measure of energy dose common in the US, but not elsewhere. As an energy dose, it does not take account of the varying biological effects of different radiation types, nor of the differing susceptibilities of the different parts of the body. 1 rad = 10 mGy.

Radiation
A particle (or ray: those terms are interchangeable) given out by some event. This book mostly concerns itself with *ionising radiation*, particularly α, β, γ and X rays and neutrons: this radiation is usually given out by processes within an atomic nucleus, but there are also other sources, such as X-ray tubes.

Radiation Accident
Any accident that might cause an additional exposure to radiation for any person. Although often very serious, they can also be relatively minor: for example, if some shielding in the X-ray department of a hospital is placed incorrectly, this is a radiation accident.

Radiation Area
An area in which increased radiation levels are allowed to occur, because they are appropriately delimited and monitored. There are two main classes: controlled areas and surveyed areas. The parts of controlled areas with the highest potential doses are defined as exclusion areas.

Radiation Exposure
The receiving of a dose from an external source, or from an incorporation. A typical level of radiation exposure from the environment is about 2.5 mSv/yr. The average exposure for a member of the public from medical diagnostics, and the use of radioisotopes in medicine and industry, is around 2 mSv/yr.

Radiation-Protection Officer
In an organisation using radioactive materials as part of its work, the radiation-protection supervisor has to appoint an appropriate number of radiation-protection officers, who will monitor the work in radiation areas. This is done formally with a radiation-protection directive. The radiation-protection officer (also called radiation officer) has to have appropriate and up-to-date qualifications for his work in the field of radiation protection. He has to organise the radiation protection, and he has to make sure that the radiation-protection regulations are respected.

Radiation-Protection Supervisor
The person responsible, at a high level, for radiation protection in an organisation. He or she has to appoint an appropriate number of radiation-protection officers by issuing a radiation-protection directive. In contrast to the radiation-protection officers, the supervisor need not be an expert in radiation protection. The radiation-protection supervisor has to ensure that the radiation-protection regulations are

respected. If there is a problem with the radiation protection in the organisation, the ultimate responsibility rests with the radiation-protection supervisor.

Radiation Sickness

This condition, also called (acute) radiation syndrome, is the result of a high radiation exposure (over 0.5 Sv). Early symptoms are nausea, vomiting and diarrhoea, followed for higher doses also by loss of hair and haemorrhage. It can be fatal, and for doses around 4 Sv, half of the exposed people would be expected to die (so 4 Sv is called the lethal dose).

Radiation Therapy

The treatment of medical conditions, particularly tumours, using methods from nuclear medicine and radiology (e.g. γ or hadron irradiation). It can also be called *radiotherapy*.

Radiation Weighting Factor

The weighting factor assessing the biological effectiveness of each type of ionising radiation. The energy dose (in Grays) multiplied by the radiation weighting factor yields the equivalent dose (in Sieverts).

Radioactive Tracer

See *nuclear tracer*.

Radioactivity

The process in which an unstable nucleus breaks down, and in doing so emits radiation (usually α, β or γ rays, but sometimes neutrons). Some types of nuclei (*radioisotopes*)

are unstable and undergo radioactive decay, and others are stable. All materials are radioactive to some small extent, due to the presence of these radioisotopes in trace quantities.

Radioisotope
An isotope which is radioactive. The radioactive properties (e.g. half-life) depend on both the number of neutrons and protons in a nucleus, but the chemical properties, and therefore the identity of the element, depend only on the number of protons. This means that different isotopes of the same element can have radically different radioactive behaviours, with some being stable, and others decaying with tiny half-lives.

Radioisotope Generator (RTG)
A Radioisotope Thermoelectric Generator produces electrical energy from the heat given out by the decay of radioisotopes.

Radiopharmaceuticals
Compounds tagged with radioisotopes, for use in clinical diagnostics. They can be administered to the patient either orally or by injections, and then used to support a scan. Tagged pharmaceuticals are also used in radiotherapy.

Radium-Beryllium Source
A neutron source. α particles from radium decay produce neutrons when they strike beryllium, according to $\alpha + {}^{9}_{4}\text{Be} \rightarrow {}^{12}_{6}\text{C} + n$.

Radon
A radioactive noble gas. The isotopes ${}^{220}\text{Rn}$ and ${}^{222}\text{Rn}$ produce about half of the natural radiation exposure of the general public (1.1 mSv/yr).

Range
The distance that a particle will travel before stopping. This depends on the particle type, the particle energy, and the properties of the material it is moving through.

Relative Biological Effectiveness (RBE Factor)
A measure of the effectiveness of a type of radiation at producing damage in a biological system. This is a complex measure, which takes into account not only the radiation type and the part of the body affected, but also the energy of the radiation, the effect being referred to, and the distribution of the dose across time. The complexity of this measure (and the associated difficulty of obtaining enough information to feed into it in individual cases) means it is not widely used.

rem
Roentgen equivalent man: an old unit for the equivalent dose. It is calculated by multiplying the energy dose in rads by the radiation weighting factor. 1 rem $=$ 10 mSv.

Restricted Area
An area to which access is limited because of possible exposure to radiation.

Risk Factor

The risk factor for a particular delayed (stochastic) radiation effect from a particular exposure is the probability that a particular individual will suffer that effect due to that exposure. For example, the leukaemia risk factor for a whole-body dose of 100 mSv is 5×10^{-4}, which means that if ten thousand people received this dose, it would be expected that 5 cases of leukaemia would be caused by it. Due to that same exposure, 11 cases of stomach cancer (risk factor 11×10^{-4}), 8 or 9 cases each of colon and lung cancer (risk factor 8.5×10^{-4} each), and about 17 cases of other cancers would also be expected, in that same group of irradiated people.

RLET

Restricted Linear Energy Transfer. When particles move through matter, they deposit more energy when moving more slowly. This means that if an incoming particle gives up a large amount of energy to an atomic electron in matter, that newly-free electron will move through the matter causing relatively little damage. By contrast, if the incoming particle gives only a little energy to the atomic electron, the electron will deposit its energy more quickly and in a more concentrated way, and so cause some damage to the material there. This means that, for radiation protection purposes, we want to ignore the collisions transferring large amounts of energy. This is what the RLET measures: the energy deposited by an incoming particle, with all collisions above a certain energy (often 100 eV) ignored. High RLET radiation, i.e. that which tends to give up small amounts of energy to large numbers of atomic electrons, is more damaging (to biological and electronic systems).

Roentgen (R)

An old measure of radiation energy deposition. It relates to another obsolete unit, the electrostatic unit (esu), which is about two billion times the charge on a single electron. One Roentgen is the intensity of X rays or γ rays which produces one electrostatic unit each of ions and electrons in 1 cm^3 of dry air. For air: 1 R = 0.88 rad = 8.8 mGy.

Scintigram

A nuclear technique to image internal organs. The patient is administered a radioisotope which will travel to certain organs. A γ-emitting isotope is chosen, such as ^{99m}Tc. These γ rays are then detected using scintillation counters, in order to image the organs. It is also possible to image certain metabolic processes in the organs using this technique.

Shielding

Any material used to protect against radiation. It is relatively easy to shield against α and β rays, because they have only short ranges in matter. A common choice to shield against γ rays is lead (any dense material is appropriate). Shielding against neutrons is best done with materials which have relatively light (low proton number) atoms, but are as dense as possible: for example water or paraffin wax. See also *absorber*.

Sievert (Sv)

The unit for the equivalent dose, which is the energy dose (in Grays), multiplied by the radiation weighting factor (different for each type of radiation). The effective (whole-body) dose, in which the different sensitivities of the different tissues are accounted for with a tissue weighting factor, is also measured in Sv.

Solid Angle
A term for the proportion of the surface of a sphere taken up by an area, by an analogy with the way an angle can be used to describe a fraction of a circle.

Spallation
A nuclear transformation caused by the impact of high-energy particles on a target, in which large numbers of nuclear fragments, α particles and nucleons are produced. This is distinct from fission, as it is the energy of the incoming particle which provides the energy for a thorough breakup of the nucleus (rather than an extra neutron inducing the splitting of an already-unstable nucleus). Spallation can be used as a high-intensity source of neutrons.

Specific Activity
The activity of a material per unit mass. This will usually be measured in Bq/kg.

Speed of Light
The speed at which light travels in a vacuum (about 1 billion kilometres per hour, or 300 million metres per second) is the fastest possible speed. No object which has mass can be accelerated to this speed, but with a large input of energy, it is possible to get very close to it. Light travels at slower speeds inside materials, so it is possible for

a particle with mass to go faster than light would inside a material, without violating this limit.

Spontaneous Fission
A type of radioactive decay in which a nucleus breaks into two smaller nuclei and some neutrons, without the trigger of an incident particle (in contrast to induced fission). It tends to occur in very heavy nuclei, with the half-life generally shorter (decay more probable) with increasing proton number.

Stochastic Radiation Effects
Delayed radiation effects for which the seriousness of the consequence does not depend on the dose, but the probability of occurrence does (and is assumed to do so linearly).

Strong Nuclear Interaction
The strongest of the three types of interaction in the nucleus (strong, electromagnetic and weak). This is the interaction, carried by particles called gluons, which binds quarks together to make protons and neutrons. It also binds together protons and neutrons to make nuclei. It is strong enough to overcome the electric force between the positively charged protons, which would act to make the nucleus fall apart. It is also called the strong force.

Subcritical
If a nuclear chain reaction has each fission causing less than one fission on average, the reaction is said to be subcritical, and the reaction will tend to slow down and stop. The term is also used for quantities of fissile material: a subcritical amount of material is too little to sustain a chain reaction on its own.

Supercritical
If a nuclear chain reaction has each fission causing more than one further fission on average, the reaction is said to be supercritical, and the reaction will tend to increase continuously. This is a very dangerous condition: if it is allowed to persist, it can lead to a meltdown or even an explosion.

Super-Heavy Hydrogen
See *tritium*.

Supervised Area
Another term for a *surveyed area*.

Surveyed Area
A radiation area less hazardous than a controlled area. According to the ICRP recommendations, individuals working in these areas might be exposed to radiation levels of more than 1 mSv per year. If the potential exposure rate is lower, the area does not have to be classified as a surveyed area.

Swimming-Pool Reactor
A nuclear reactor in which the core is immersed in a large, open tank of water. The water glows blue due to Cherenkov radiation from fast-moving electrons. These fast-moving electrons originate from nuclear β decay of some fission products.

Synchrotron

A synchrotron accelerates particles on a circular course through an evacuated pipe. As the particles are accelerated, the guiding field must be increased in order to continue to bend the particles' paths sufficiently to keep them moving around the circle. As their paths are bent, the particles emit X rays. These X rays have a wide variety of scientific and technical applications, including determining the structure of biological molecules. Currently-operational examples of synchrotrons include the Advanced Photon Source at the Argonne National Laboratory in the US, and the European Synchrotron Radiation Facility near Grenoble in France.

Terrestrial Radiation

Natural ionising radiation from the Earth's crust. Typical radioisotopes are potassium-40, radium-226 and thorium-232. The radiation exposure from terrestrial radiation in most parts of the world is about 0.5 mSv/yr, but it is significantly higher in some places.

Tissue Weighting Factor

The weighting factor assessing the different radiation sensitivities of different tissues and organs. The ICRP-recommended tissue weighting factors are: red bone marrow 0.12, stomach 0.12, colon 0.12, lung 0.12, chest 0.12, gonads 0.08, bladder 0.04,

Deinococcus radiodurans and deinococcus radiophilus enjoy
splashing around in the core of a swimming-pool reactor!

oesophagus 0.04, liver 0.04, thyroid gland 0.04, periosteum (bone surface) 0.01, skin 0.01, brain 0.01, salivary glands 0.01, other organs or tissue 0.12.

Tokamak

A toroidal (doughnut-shaped) combustion chamber within magnetic coils, in which a deuterium-tritium plasma is fused. At the time of writing, it has not yet been possible to gain more energy out of a tokamak than was put in to create and maintain the plasma, but there is active research into this. The name originates from an abbreviation in Russian.

Transmutation

Long-lived radioactive waste can in principle be transformed into short-lived or even stable isotopes by proton or neutron bombardment. This technique has so far only been demonstrated to work on small samples, but research is continuing.

Tritium

A radioactive isotope of hydrogen with two neutrons (^{3}H), sometimes also called super-heavy hydrogen. It can be written with the symbol T, so T_2O is water with two tritium atoms instead of normal hydrogen. It is common in nuclear physics to write a nucleus of tritium (called a *triton*) as t. The activity of normal drinking water due to tritium is around 0.1 Becquerels per litre.

Uranium

The most significant natural radioactive element. In nature, it is a mixture of isotopes, with mostly ^{238}U (half-life 4.5 billion years), and about 0.7 % ^{235}U (half-life 0.7 billion years). It is the ^{235}U which is fissile, and is therefore the component most useful in power plants and weaponry. Most radioactive material for industrial and military uses is derived ultimately from uranium.

Valence Band

The *band theory* describes the energy levels in a solid, saying that there are bands in which electrons can sit. The lowest-energy bands will be fully filled. Of these fully-filled bands, the highest-energy one is called the valence band. The band directly above it in energy will be the *conduction band*. Radiation will sometimes promote electrons from the valence band to the conduction band, where they can then conduct electricity.

Weak Nuclear Interaction

The weakest of the three types of interaction in the nucleus (strong, electromagnetic and weak). This is the interaction, carried by particles called W and Z bosons, which allows β decay to take place, transforming a neutron into a proton and releasing an electron and an antineutrino, or vice versa (with a positron released). It is also called the weak force.

Weighting Factor

A measure of the effectiveness of a type of radiation to produce a biological effect. There are two types of weighting factors applied to an energy dose to get a realistic understanding of its effect: see *radiation weighting factor* and *tissue weighting factor*.

X Rays
Electromagnetic radiation with energies between $100\,eV$ and $100\,keV$ (these boundaries are defined in different ways by different authors). X rays are created by transitions between inner shells of atoms, by bremsstrahlung, and by synchrotrons. X rays are mainly used for diagnostic radiography and for crystallography. X rays are a form of ionising radiation.

Zinc Sulphide
A chemical, also called 'zinc blende', which can be used to construct a scintillation screen, i.e. a screen which emits scintillation light upon impact of charged nuclear particles or γ rays. Zinc sulphide screens have been in use since the very early days of work on radioactivity.

Index

© Springer International Publishing Switzerland 2016
C. Grupen and M. Rodgers, *Radioactivity and Radiation*,
DOI 10.1007/978-3-319-42330-2